Artificial Intelligence with Chinese Characteristics

Jinghan Zeng

Artificial Intelligence with Chinese Characteristics

National Strategy, Security and Authoritarian Governance

palgrave
macmillan

Jinghan Zeng
Lancaster University
Lancaster, UK

ISBN 978-981-19-0721-0   ISBN 978-981-19-0722-7   (eBook)
https://doi.org/10.1007/978-981-19-0722-7

Cover credit : Andriy Onufriyenko/GettyImages

This Palgrave Macmillan imprint is published by the registered company Springer Nature
Singapore Pte Ltd.
The registered company address is: 152 Beach Road, #21-01/04 Gateway East, Singapore
189721, Singapore

ACKNOWLEDGEMENTS

This book is dedicated to Emma, Lina and Luna who are the miracles that make my life complete and meaningful. This book manuscript was completed during an unexpected journey in China. Without their support and love, this book would not be possible.

I thank the respective publisher for granting permission to reproduce some materials of my previously published articles including

- Jinghan Zeng, China's Date with Big Data: Will It Strengthen or Undermine the Authoritarian Rule? *International Affairs*, Vol. 92, No. 6, November 2016, 1443–1462
- Jinghan Zeng, Artificial Intelligence and China's Authoritarian Governance, *International Affairs*, Vol. 96, No. 6, November 2020, 1441–1459
- Jinghan Zeng, China's Artificial Intelligence Innovation: A Top-Down National Command Approach? *Global Policy*, Vol. 12, No. 3, May 2021, 399–409
- Jinghan Zeng, Securitization of Artificial Intelligence in China, *Chinese Journal of International Politics*, Vol. 14, No. 3, Autumn 2021, 417–445

I greatly appreciate all reviewers for their insightful comments on my book manuscript and the above articles. I am also grateful to Jonna

Nyman, Tim Stevens and Yuefan Xiao for their inspiration and valuable comments during the writing of these articles.

CONTENTS

1 Introduction   1

2 China's AI Approach: A Top-Down Nationally
  Concerted Strategy?   11

3 China's Security Politics of AI   35

4 China's Authoritarian Governance and AI   67

5 Towards a Research Agenda for AI with Chinese
  Characteristics   105

References   109

ABBREVIATIONS

AI	Artificial Intelligence
BRI	Belt and Road Initiative
BRICS	Brazil, Russia, India, China and South Africa
CCP	Chinese Communist Party
EU	European Union
ICT	Information and Communication Technology
US	United States

LIST OF FIGURES

Fig. 2.1   Summary of Chinese provincial-level unit target values
           for the AI core industry by 2020 (unit: 1 billion yuan)   22
Fig. 2.2   Summary of Chinese provincial-level unit target values
           of the AI-related industry by 2020 (unit: 1 billion yuan)   23
Fig. 3.1   The number of Chinese academic articles with the words
           "AI" and "security" in the title in China's CNKI database
           (2001–2019)   53

CHAPTER 1

Introduction

The artificial intelligence (AI) revolution is likely to bring about significant changes to human society for better or worse. Despite concerns about the negative impact that AI may bring, China has decided to openly embrace the age of AI and accelerate its arrival. In 2017, the Chinese government openly spoke about its global ambition to become a leading AI power by 2030 (China 2015).China's open embracing of the age of AI has attracted considerable policy, academic and media attention. In an era of great power politics, China's AI dream, led by its economic rise, has heated up tech competition on the global stage. To many, Beijing's bold AI plans open a new front for US-China geopolitical competition (Klein 2020; Allison 2019; Castro et al. 2019; Allen 2019; Kempe 2019). The discourse of an "AI race" or "AI cold war" has made for many headlines. With a few exceptions,[1] the Chinese approach to AI development is frequently labelled as a "top-down" national strategy (Liu 2019; Ramanathan 2019; Ives and Holzmann 2018; Jing 2018; Yu 2019; Lanier and Weyl 2020; Lee 2018). To many, China's AI strategy is some sort of "unified/integrated" "whole-of-nation/government/society" approach to drive AI innovation (Ding 2018; Kania 2017; Hoadley and Sayler 2020). This "unique" Chinese approach and its underlying strategic

---

[1] For example, Ding argues that China's AI strategy is "not a monolithic, completely top-down approach; many actors are maximizing their own interests and responding to broad signals from the central government" (Ding 2018: 20).

© The Author(s), under exclusive license to Springer Nature Singapore Pte Ltd. 2022
J. Zeng, *Artificial Intelligence with Chinese Characteristics*,
https://doi.org/10.1007/978-981-19-0722-7_1

thinking has led to a debate over whether China is winning the AI race, causing considerable anxiety for China's near-peer competitors, especially the US (Allison 2019; Allen 2019; Kempe 2019).

Indeed, this AI competition is not only technological but also ideological, as how to govern the use of AI is guided by political values. A key goal of the "American AI Initiative"—the United States' (US) national strategy for maintaining American leadership in AI, launched in 2019—was to ensure that American AI innovation was underpinned by American values. It aimed to develop AI technologies reflecting fundamental American values such as freedom, guarantees of human rights, the rule of law and rights to privacy (US 2019). In Europe, an ethics-first approach has been adopted by the European Union (EU) (Roberts et al. 2020) to strictly restrict the use of AI (Satariano 2021). In comparison, China's bold AI practices have been underpinned by its authoritarian values. The authoritarian regime has actively made use of AI and big data-related technologies to improve its governance. China's digital progress benefits from its huge internet market, strong state power and weak civil awareness, making it more competitive than Western democratic societies where privacy concern restricts their AI development. Thus, according to Webb (2019a), "Beijing's AI push is part of a coordinated attempt by President Xi Jinping to turn China into the world's unchallenged AI hegemon, and to control and monitor its citizens…the US – working with its democratic partners – urgently needs to play catch up and develop the strong, solid muscles it will need to win the AI race".[2]

Not surprisingly, China is often labelled as the US's "most serious strategic competitor" (NSCAI 2019: 6) or "closest competitor" (Hoadley and Sayler 2020: 21) in American AI policy discourses and thus becomes an inevitable topic and useful reference point for American AI strategy. As a US Congress report on AI and National Security points out:

> Most analysts view China's unified, whole-of-government effort to develop AI as having a distinct advantage over the United States' AI efforts. (Hoadley and Sayler 2020: 24)

---

[2] Similarly, in Europe, when highlighting the necessity of upholding liberal and democratic norms in AI algorithm, some suggest that "we infuse technology with values all the time, which obviously leads to systematic problems when it comes to China" (Mueller-Kaler 2020).

As such, many are calling for a similar "nationally concerted" and "whole-of-government" approach to AI in the US.

For American tech giants, an exaggerated China threat is also helpful in fending off criticism towards their monopoly by appealing to American national interests. Market actors in the US have taken advantage of this China threat for their cause. For example, when facing pressure to break up Facebook, China is referenced to justify Facebook's monopoly. Facebook CEO Mark Zuckerberg once warned that Chinese tech companies would dominate if Facebook was unravelled. According to Zuckerberg,

> if the (US) government here is worried about – whether it's election interference or terrorism – I don't think Chinese companies are going to want to cooperate as much and aid the national interest there. (Wagner 2018)

Here, Zuckerberg labelled Chinese tech companies as uncooperative foreign entities that would undermine key American national interests, using election interference and terrorism as reference examples.

Consequently, in the relevant policy discourses, AI becomes a matter of security, and China's—and Russia's—AI advancement is often labelled as a threat to national and international security. By framing AI a national security matter rather than a normal technology, it justifies the need to enable extraordinary actions from the state and society—in the American case, for example, to deploy more resources and support to not only the American AI-enabled military sector but also the AI commercial industry. Likewise, China is using the US as a mirror to reflect what it should do in this AI competition, and there is also a growing tendency to label AI as a security matter in the Chinese policy discourses.

Key Arguments of the Book

The above context involves three sets of interrelated aspects about China's AI politics. The first is how to characterize China's AI strategy. Is it a top-down national command approach, as many have argued? Is it geopolitically driven by a coherent, consistent and unified central objective set by Beijing, as many have suggested? Which domestic and international actors have decided and implemented China's AI strategy? Have they made a nationally concerted effort to achieve Beijing's goals of pursuing a China-centred AI order, as many have suggested? The second set of questions is about security and China's policy discourse of AI. How has China's policy discourse described AI? Why is the Chinese government labelling AI as a matter of security? What is the impact of this security discourse on China's AI policy objectives? The third set of questions is about China's authoritarian governance and AI. How has AI been empowering China's authoritarian governance? What are the key opportunities and challenges brought about by China's bold AI practices?

This book aims to answer these questions. The overall argument is that China's AI approach is sophisticated and multifaceted. This AI with Chinese characteristics approach is shaped by its domestic politics and has brought about both benefits and challenges to China. More specifically, this book first argues that China's AI plans are primarily driven by contestation and the struggle for resources among domestic stakeholders who are economically motivated and have little awareness of the bigger geopolitical picture. In addition, China's national AI plan is an upgrade of existing local AI initiatives to the national level, reflecting a bottom-up development. As such, China's AI approach is not a top-down geopolitically driven nationally concerted strategy. In this regard, the existing analyses vastly exaggerate (a) Beijing's capacity to coordinate domestic capital and actors towards a unified, specific strategic objective and (b) the extent of China's AI advancement and its geopolitical threat, triggering unnecessary anxiety among China's near competitors. China's AI strategy is thus better understood as a broad policy manifesto, accommodating competing interests of various domestic stakeholders.

Moreover, related to the point above, precisely because of the significance of China's subnational and non-state actors in China's AI politics, the Chinese government is looking for ways to mobilize domestic actors. This book argues that AI is being securitized by the Chinese central government to mobilize local states, market actors, intellectuals and the general public. China's historical anxieties about technology and regime security needs are contributing to the rise of security discourse in China's AI politics, and this securitization trend is further accelerated by the growing tension caused by great power competition. Despite its help in convincing domestic actors, however, this securitization trend may undermine Chinese key AI objectives by pushing it in an inward-looking, techno-nationalistic direction that brings about a series of severe consequences for China's AI industry and leadership ambition.

Lastly, the Chinese government has heavily invested in AI's potential in governance. This book argues that China's bold AI practices are part of its broad and incoherent adaptation strategy to governance by digital means.

AI is part of a digital technology package that the Chinese authoritarian regime has actively employed not only to improve public services but also to strengthen its authoritarian governance. While China's AI progress benefits from its unique political and social environment, its ambitious AI plan contains considerable risks. The overall impact depends on how AI affects major sources of political legitimacy, including economic growth, social stability and ideology. China's approach is gambling on its success in (a) delivering a booming AI economy, (b) ensuring a smooth social transformation to the age of AI, and (c) proving ideological superiority of its authoritarian and communist values.

Above all, as the first book-length academic study of China's AI politics, this book aims to advance understanding of AI with Chinese characteristics, which is crucial to inform what lessons we can learn from the Chinese approach to AI and how to respond to China's rise as a global AI leader.

DEFINING THE BUZZWORDS

This book predominately studies AI but it also touches upon big data-related technologies—especially in Chapter 4 where the authoritarian use of digital technologies is discussed. Both AI and big data are the buzzwords the world over, including China. Both of them are umbrella terms that refer to many meanings in different contexts. In this book, AI refers to a wide range of digital technologies with the ability "to perform tasks that would usually require human intelligence" (Oxford 2005). AI is different but closely related to big data. The latter is primarily about data that is collected and AI algorithms rely heavily on data for training and development. While big data may include traditional (non-digital) sources of data, this book will primarily focus on digital data sources. Thus, big data is broadly defined as an "explosion in the quantity and diversity of high frequency digital data" in this book (UN 2012). Given data's necessity in training AI algorithms, greater access to big data puts China in an advantageous position in the global race.

In China, both AI and big data are not uniform concepts, and are sometimes used as catch phrases. It is also important to draw a distinction between big data as a phenomenon and the way attempts are made to use those data. In China, slogans such as "to promote the development of big data" or "to develop big data" frequently appears in official documents. However, these words refer to improvements in the way that big data is used and applied rather than to develop big data as a phenomenon. In addition, "to vigorously develop AI" and "to vigorously develop big data" have been used by the state as policy slogans to mobilize domestic actors, leaving considerable room for interpretation. Thus, there is neither coherent understanding nor unified use of these two concepts in China.

Notably, there are three types of AI: narrow, general and super (Kaplan and Haenlein 2019). Narrow AI (also called weak AI) refers to digital technologies with a narrow range of ability that is dedicated to specific tasks such as the iPhone virtual assistant Siri, drone robots and self-driving cars. It is the most basic generation of AI and performs below human level. The current state of AI development belongs to this narrow AI generation. General AI (also called strong AI) is a more advanced generation that has yet to be achieved; it will have cognitive abilities that can perform as well as human intelligence. Super AI—which is still hypothetical—represents the most advanced generation of digital technology currently conceived; it will have strong self-awareness and be able to surpass human intelligence in all areas. This book is mostly about narrow AI, but also briefly touches upon general and super AI.

BOOK OUTLINE

This chapter introduces the content of the book including the background, current discussion about China's AI politics, key arguments and book structure.

Chapter 2 studies the characterization of China's AI strategy and its domestic contestation. It argues that the existing analyses about China's AI strategy are mistaken. By examining the nature of China's fragmented authoritarian system and the role of China's state and non-state actors in the AI-related policy process, this chapter shows why China's AI approach is not a top-down nationally concerted command approach in order to achieve Beijing's geopolitical goals.

Chapter 3 analyses China's security politics of AI. By using securitization as a conceptual framework, it shows how and why the Chinese central government is securitizing AI in order to mobilize domestic actors. It argues that this security discourse is not only underpinned by China's historical anxieties about technology and regime security needs but also accelerated by the growing geopolitical tensions. This chapter also suggests that this securitization trend may bring about unintended consequences for China's AI objectives.

Chapter 4 studies China's governance approach to AI and big data-related technologies. It discusses how and why China's authoritarian regime has taken a proactive approach to invest in AI's potential for governance. Despite the immediate benefits offered by AI and big data-related technology, China's ambitious approach is facing considerable challenges. This chapter argues that how to (a) deliver strong growth in China's AI economy, (b) handle the dramatic social transformation to the age of AI and (c) justify legitimacy of its political values embedded in its AI approach represent three critical tests for China's governance approach to AI.

Chapter 5 summarizes the key arguments of the book and suggests a future research agenda for AI with Chinese characteristics.

References

Allen, Gregory. 2019. *Understanding China's AI Strategy*. Center for a New American Security. Available at https://www.cnas.org/publications/reports/understanding-chinas-ai-strategy. Accessed on 28 Feburary 2020.

Allison, Graham. 2019. "Is China Beating America to AI Supremacy?" *The National Interest*. Available at https://nationalinterest.org/feature/china-beating-america-ai-supremacy-106861. Accessed on 28 Feburary 2020.

Castro, Daniel, Michael McLaughlin, and Eline Chivot. 2019. *Who Is Winning the AI Race: China, the EU or the United States?* Center for Data Innovation. Available at https://www.datainnovation.org/2019/08/who-is-winning-the-ai-race-china-the-eu-or-the-united-states/. Accessed on 28 Feburary 2020.

China. 2015. "国务院关于印发促进大数据发展行动纲要的通知 (State Council's Decision on Promoting the Development of Big Data)." Available at http://www.gov.cn/zhengce/content/2015-09/05/content_10137.htm. Accessed on 20 November 2015.

Ding, Jeffrey. 2018. *Deciphering China's AI Dream*. Future of Humanity Institute, University of Oxford. Available at https://www.fhi.ox.ac.uk/deciphering-chinas-ai-dream/. Accessed on 28 Feburary 2020.

Hoadley, Daniel, and Kelley Sayler. 2020. *Artificial Intelligence and National Security*. Congressional Research Service Report. Available at https://fas.org/sgp/crs/natsec/R45178.pdf. Accessed on 3 January 2021.

Ives, Jaqueline, and Anna Holzmann. 2018. *Local Governments Power Up to Advance China's National AI Agenda*. Mercator Institute for China Studies. Available at https://www.merics.org/en/blog/local-governments-power-advance-chinas-national-ai-agenda. Accessed on 28 Feburary 2020.

Jing, Meng. 2018. *Is Xi Jinping's Iron Grip Better Than Adam Smith's Invisible Hand for Technology Innovation?* Available at https://www.scmp.com/tech/

article/2173128/xi-jinpings-iron-grip-better-adam-smiths-invisible-hand-tec hnology-innovation. Accessed on 28 Feburary 2020.

Kania, Elsa. 2017. "China's Artificial Intelligence Revolution: A New AI Development Plan Calls for China to Become the World Leader in the Field by 2030." *The Diplomat*.

Kaplan, Andreas, and Michael Haenlein. 2019. "Siri, Siri, in My Hand: Who's the Fairest in the Land? On the Interpretations, Illustrations, and Implications of Artificial Intelligence." *Business Horizons* 62: 15–25.

Kempe, Frederick. 2019. "The US Is Falling Behind China in Crucial Race for AI Dominance." CNBC. Available at https://www.cnbc.com/2019/01/25/ chinas-upper-hand-in-ai-race-could-be-a-devastating-blow-to-the-west.html. Accessed on 28 Feburary 2020.

Klein, Andrés Ortega. 2020. *The U.S.-China Race and the Fate of Transatlantic Relations Part 1: Tech, Values, and Competition*. The Center for Strategic and International Studies (CSIS). Available at https://www.csis.org/analysis/us-china-race-and-fate-transatlantic-relations. Accessed on 28 Feburary 2020.

Lanier, Jaron, and E. Glen Weyl. 2020. "How Civic Technology Can Help Stop a Pandemic." *Foreign Affairs*. Available at https://www.foreignaffairs.com/ articles/asia/2020-03-20/how-civic-technology-can-help-stop-pandemic. Accessed on 20 March 2020.

Lee, Kaifu. 2018. *AI Superpowers: China, Silicon Valley, and the New World Order*. Houghton Mifflin Harcourt.

Liu, Yiling. 2019. "China's AI Dreams Aren't for Everyone." *Foreign Policy*. Available at https://foreignpolicy.com/2019/08/13/china-artificial-intellige nce-dreams-arent-for-everyone-data-privacy-economic-inequality/. Accessed on 28 Feburary 2020.

Mueller-Kaler, Julian. 2020. *Europe's Third Way*. Atlantic Council. Available at https://www.atlanticcouncil.org/content-series/smart-partnerships/ europes-third-way/. Accessed on 16 September 2021.

NSCAI. 2019. *The Interim Report of National Security Commission on Artificial Intelligence*. National Security Commission on Artificial Intelligence. Available at https://drive.google.com/a/nscai.org/file/d/153OrxnuGEjsUvIxWs FYauslwNeCEkvUb/view?usp=sharing. Accessed on 21 June 2020.

Oxford. 2005. Artificial Intelligence. In *The Oxford Dictionary of Phrase and Fable*. Available at https://www.oxfordreference.com/view/10.1093/oi/aut hority.20110803095426960. Accessed on 18 June 2020: Oxford University Press.

Ramanathan, Shriram. 2019. *China's Booming AI Industry: What You Need to Know*. Lux Research. Available at https://www.luxresearchinc.com/blog/chi nas-booming-ai-industry-what-you-need-to-know. Accessed on 28 Feburary 2020.

Roberts, Huw, Josh Cowls, Emmie Hine, Jessica Morley, and Mariarosaria Taddeo. 2020. "Governing Artificial Intelligence in China and the European Union: Comparing Aims and Promoting Ethical Outcomes." Available at SSRN: https://ssrn.com/abstract=3811034.

Satariano, Adam. 2021. "Europe Proposes Strict Rules for Artificial Intelligence." *The New York Times.*

UN. 2012. *Big Data for Development: Opportunities and Challenges.* Available at http://www.unglobalpulse.org/sites/default/files/BigDataforDevelopment-UNGlobalPulseJune2012.pdf. Accessed on 15 February 2016: UN Global Pulse.

US. 2019. *Artificial Intelligence for the American People.* The White House. Available at https://www.whitehouse.gov/ai/. Accessed on 3 January 2021.

Wagner, Kurt. 2018. "Mark Zuckerberg Says Breaking Up Facebook Would Pave the Way for China's Tech Companies to Dominate." *Vox Media.*

Webb, Amy. 2019. "Build Democracy into AI: Human-Centered Policy is Needed to Wrest Control from China, Tech Giants." *Politico.*

Yu, Yifan. 2019. "Why China's AI Players Are Struggling to Evolve Beyond surveillance." *Nikkei Asian Review.* Available at https://asia.nikkei.com/Spotlight/Cover-Story/Why-China-s-AI-players-are-struggling-to-evolve-beyond-surveillance. Accessed on 1 March 2020.

CHAPTER 2

China's AI Approach: A Top-Down Nationally Concerted Strategy?

INTRODUCTION

China's open ambition of becoming a global AI superpower by 2030 has attracted considerable attention, as mentioned in Chapter 1. Many argue that China has taken advantage of its national approach to contest for AI supremacy and geopolitical dominance. The relevant analyses often characterize China's AI plans as being Beijing's coherent top-down strategy. As Ives and Holzmann (2018) put it,

> The current top-down approach in China's AI industry is thus in line with the country's overall industrial policy, in that it mobilizes massive amounts of capital and labor towards a specific target even at the risk of creating inefficiencies and wasting resources.

Chinese and American approaches to AI are often used for comparison. Many relevant analyses consist of a simple dichotomy: China's top-down nationally coordinated approach versus the American market-oriented approach.[1] As Amy Webb (2019) argues,

---

[1] For example, Yu (2019) argues that "American AI development is market-driven and dominated by independent, private sector players … By contrast, China's approach is centralized and top down". Similarly, many characterize Chinese and American approaches as "state-focus" and "company-focus" models, respectively when discussing the role of the

J. Zeng, *Artificial Intelligence with Chinese Characteristics*, https://doi.org/10.1007/978-981-19-0722-7_2

The future of AI is currently moving along two developmental tracks that are often at odds with what's best for humanity. China's AI push is part of a coordinated attempt to create a new world order led by President Xi, while market forces and consumerism are the primary drivers in America.

In addition, China's AI strategy is also considered by many as some sort of "unified/integrated" "whole-of-nation/government/society" approach, which has given China considerable advantages over its peer competitors' AI efforts, as mentioned in Chapter 1. In this regard, the Chinese approach is often summarized as a "geopolitically driven" "top-down" national strategy reflecting the ambition of Beijing to pursue a China-centred AI order and assuming a concerted national effort to achieve a unified central objective.

This chapter, however, argues that these views are mistaken. It argues that, in order to mobilize domestic actors, China's AI plans are kept deliberately vague and broad to accommodate the interests of domestic stakeholders. Instead of unfolding according to Beijing's top-level design, China's AI development is primarily driven by powerful domestic stakeholders with diverse and competing interests. As economic growth is the most important goal of China's AI plans, the central state has restricted discretion, while local states have primary responsibility for boosting the AI economy in China. Although this division of authority provides institutional incentives for local states to promote the AI industry in their own jurisdictions, it creates additional difficulties of steering and coordination. As this chapter will show, the diverse interests among domestic actors have produced a high level of regional competition regarding the AI industry.

This game of contestation, and the struggle for resources, has been played by local, subnational and non-state actors who are economically motivated with little awareness of diplomatic or geopolitical goals. This not only disproves the "coordinated" narrative in terms of Chinese AI strategy but also shows that global geopolitical factors are somewhat irrelevant at the local level. In addition, local provinces and non-state actors had already made their AI plans long before the central government announced AI as a national strategy in 2017 (Ding 2018: 15). The latter was indeed a recognition and upgrade of existing successful

EU in the global AI race and how the EU can find a "third way" (Mueller-Kaler 2020; Burrows and Mueller-Kaler 2021).

initiatives at the local level, reflecting a bottom-up rather than a top-down development.[2] The rapid development of China's AI industry is largely delivered by China's market forces, especially proactive profit-driven private entrepreneurship, rather than the instructions of central agencies in Beijing.

This chapter points to complex and multidimensional China's AI politics. In the relevant AI race analyses, the geopolitical ambition of China's AI strategy has often been highlighted. Using cold war zero-sum thinking to analyse China's AI innovation, these arguments have inevitably triggered unnecessary anxiety for China's near-peer competitors. By arguing a distinct advantage of this "unified" Chinese masterplan, many analysts urge immediate policy actions—for example, the US government to adopt a similar approach in order to maintain its AI supremacy (Corrigan 2018). These arguments vastly exaggerate Beijing's capacity to coordinate domestic capital and actors towards a unified, specific strategic objective.

In this regard, China's AI innovation is much less threatening to China's near AI competitors in geopolitical terms than it is framed in the relevant analyses. Given China's important role in AI innovation, the exaggeration of a China threat is unhelpful to transnational cooperation in AI that can benefit the world. For policy-makers of China's near AI competitors such as the US and Europe, it is critical to make evidence-informed decisions that take these complex and multifaceted domestic dynamics of China's AI development into consideration where comparisons are made.

This chapter proceeds in four parts. The first part discusses the transformation of China's governance model over the past few decades, and why its operation mechanism is not a top-down command approach. The second part analyses how this transformed governance model operates regarding China's AI plans and how the state AI strategy follows a bottom-up development. It also shows that domestic (non-)state actors have taken advantage of the central government's AI plans to maximize their own interests, exposing the problems of steering and coordination. The third part explains why the Chinese central state chooses not to

---

[2] In this book, a bottom-up approach refers to the process of policy ideas developed from the bottom level of the hierarchy, i.e. local states upwards to the top level, i.e. the central government.

intervene to a larger extent and discusses its own coordination problems. The conclusion summarizes the relevant arguments and discusses the implications for China's AI development.

FEDERALISM WITH CHINESE CHARACTERISTICS

Within the literature of international relations, the analyses of China often assume that China/the Chinese state is a monolithic political entity. The Chinese authoritarian regime is simply considered a unified and "highly centralized" decision-making system, which Beijing can easily mobilize to achieve its goals (Hill 2016). This simplified understanding inevitably neglects the nuanced development of China's domestic political economy and its impacts on China's foreign relations. In fact, decades of scholarship in China Studies have produced a sizeable literature that presents a fragmented and decentralized authoritarian system in China (Lieberthal and Oksenberg 1988; Lieberthal 1992; Lieberthal and Lampton 1992; Schurmann 1966).

Since the early 1980s, China's party state has been transformed by its market reforms. In contrast with public understanding, this transformation has led to a series of political reforms in China—despite not for the sake of democratization (Zeng 2015b). Driven by a pragmatic and market-oriented approach, China's rapid economic changes have altered its central-local relations. These reforms programmes have introduced a division of authority between central and local states (Montinola et al. 1996; Xu 2011). This has given local states primary responsibility for regional economic prosperity at the cost of the central government's monopoly control over the Chinese economy. Consequently, this has led to a high level of local autonomy and a restricted discretion of the central authority (Montinola et al. 1996; Xu 2011). As Shaun Breslin points out, market reforms programmes have allowed local leaders not only the ability to "ignore central economic commands" but also "the desire to assert their independence" (Breslin 1996).

This federalism without a separation of power and popular elections is known as "federalism, Chinese style" (Montinola et al. 1996; Qian and Weingast 1995), "de facto federalism" (Zheng 2007) or federalism with Chinese characteristics. The benefits of this fiscal decentralization are obvious. The most notable one is giving local actors institutional incentives to promote economic growth. Many consider this fiscal decentralization the key to China's economic miracles over the past decades

(Qian and Weingast 1995; Montinola et al. 1996). This system gives local states the power and motivation to explore the most efficient economic policies and thus boost regional growth.

Restricted central authority and local independence, however, inevitably increase the problems of steering and coordination. This transformation has altered the power relations and the bureaucratic order by encouraging a high level of competition among regional actors (Montinola et al. 1996; Xu 2011).In addition to fighting for the central government's support, this competition extends to the struggle for factors of production including labour and capital (Montinola et al.1996). As a result, regional relations have become increasingly competitive. Local fiscal independence also means that local actors are primarily driven by their individual interests rather than China's geopolitical goals.

In addition, fiscal decentralization has fundamentally changed the way the central state attains its objectives. The decision-making process is no longer a simple top-down command approach imposed by a few top leaders in Beijing, if it ever was. It now involves "ongoing, multi-level, multi-agency bargaining, whereby apparently subordinate actors may influence, interpret or even ignore central policy" (Jones and Zeng 2019: 1416). In this regard, policy slogans are crucial for central-local interaction. In order to mobilize domestic actors, the central state has to rely on policy slogans that are often vague and loose to leave room for local actors to adapt them to local conditions (Zeng 2020). The process of translating these vague slogans into practices will inevitably invite agendas and interests of local actors (Zeng 2019, 2020). When economic interests are involved, local actors will compete to jump on the bandwagon to fight for support from the central government, creating the illusion of a "whole-of-nation" approach to make concerted efforts towards achieving the central goals while hiding the high level of regional competition and the manipulation of policy slogans for individual/regional gains (Zeng 2020).

This Chinese federalism also means that decision-making is not always a one-way process of a top-down translation. Instead, decentralized policy experimentation and bottom-up initiatives represent a key part of the central-local interaction (Zheng 2007). This policy experimentation allows local states to test or explore novel policy ideas and thus promotes policy innovation or institutional changes. When successful local policy initiatives are developed, the central government may choose to adopt those local initiatives at the national level. Thus, policy experimentation is a process of turning local initiatives into national policies. As Heilmann (2008b: 21) argues, it has "minimized the risks and the cost to central policymakers by placing the burden on local governments and providing welcome scapegoats in cases of failure".

Indeed, many Chinese national policies are developed from local initiatives via this bottom-up manner (Heilmann 2008b, 2018). This has developed a sizeable literature of policy experimentation (Heilmann 2008b, 2018, 2008a, 2009; Zeng 2015a; Wang 2009; Parris 1993). Take China's national talent policy as an example. It was developed from local talent policies in cities such as Wuxi that aimed to attract oversea talent to grow their business in Wuxi as returnee entrepreneurs to boost the city's innovation and economy (Xing et al. 2018). In order to maintain their regional competitiveness, other cities and provinces have emulated and developed this talent policy based on their local needs, leading to different forms of talent policies at the regional level. Given the policy success at the local level, the central government has recognized these local initiatives and institutionalized talent policies at the national level in order to promote China's innovation and national economy.

Unfortunately, the above "conventional wisdom" within the China scholarship has not changed the way that China is understood by the community of international relations despite efforts made by many China experts. To some extent, this is understandable given that China's rise is a recent phenomenon, and, after all, this knowledge is traditionally considered "domestic". Yet, the line between domestic and international Chinese politics has become increasingly blurred nowadays. The transformation of China's authoritarian system has been further deepened by globalization and China's economic integration with the world, granting domestic actors growing transnational interests and global influence. Provincial governments that previously had small international roles have now become powerful stakeholders managing transboundary economic and security matters (Su 2015) and exerting their global influence via, for example, associated state-owned enterprises (Jones and Zou 2017). As such, the Chinese authoritarian system has become not only increasingly fragmented and decentralized but also "internationalized" (Jones and Zeng 2019; Hameiri and Zeng 2020; Hameiri and Jones 2016; Jones 2019; Hameiri et al. 2019).

In this context, the domestic technocratic problems of steering and coordination have produced wider and deeper global consequences. The knowledge gap between International Relations and China Studies has become increasingly negatively impactful. The Belt and Road Initiative (BRI) is a recent notable example. The expansion of China's BRI has received extensive academic and public attention given the enormous resources involved. Following a traditional top-down monolithic approach, many argue that the BRI is Beijing's "well-thought-out" and "clearly defined" grand strategy to reshape global order and achieve geopolitical dominance (Callahan 2016; Miller 2017; Leverett and Wu 2017). Yet, as Jones and Zeng argue, the design and implementation of the BRI is actually shaped by a bottom-up manner driven by local, subnational actors who struggle for resources with little awareness of diplomatic and geopolitical goals (Jones and Zeng 2019). Driven by the diverse and competing interests of powerful domestic and international stakeholders, the BRI has been unfolding in an incoherent and fragmented fashion rather than as some grand strategic masterplan designed in Beijing (Jones and Zeng 2019). This means that it may depart significantly from the top-level plan made by the Chinese central government (Jones and Zeng 2019).

Similarly, the diverse and competing domestic interests have led to contradictory Chinese engagement on the global stage from the South China Sea (Group 2012; Hameiri and Jones 2016) to even within the hardest security arenas, such as nuclear governance (Hameiri and Zeng 2020). Instead of shaping the world on Beijing's terms, the outcomes of these domestic dynamics sometimes undermine Chinese foreign policy objectives (Jones and Zeng 2019). So, how will this transformed party-state system inform our understanding of China's AI development? The following section answers this question.

CHINA'S AI ADVANCEMENT: A TOP-DOWN COMMAND APPROACH?

China's AI plans started to attract global attention mainly during the period from 2016 to 2017 when the Chinese central government announced a series of AI strategy papers. In May 2016, the "'Internet + ' AI Three Years Implementation Plan" was jointly put forward by the National Development and Reform Commission, the Ministry of Science and Technology, the Ministry of Industry and Information Technology and the then Central Leading Group for Cyberspace Affairs. In July 2017, the Chinese State Council put a further national focus on AI and issued the "New Generation AI Development Plan".

This document specified China's three-step AI plan with quantitative targets (China 2015). The first step in China's overall AI strategy is to reach the same level as leading countries such as the US and develop an AI industry worth more than 150 billion yuan by 2020, followed by the second step of making major breakthroughs in some parts of AI technology and developing an industry worth more than 400 billion yuan by 2025. The third and last step is to become the leading AI power with an industry worth more than 1000 billion yuan by 2030 (China 2015). Following this, in October 2017, the 19th Chinese Communist Party (CCP) report emphasized the importance of promoting the development of AI in order to accelerate the process of making China a major manufacturing power (Xi 2017). Afterwards, the concept of AI frequently appeared in various governmental plans and reports.

As mentioned previously, the high-profile Chinese state announcement of AI as a national strategy has produced many international analyses focusing on Beijing's strategic thinking regarding the AI race and its geopolitical goals. Many of these arguments are valid. As Xi Jinping points out, "accelerating the development of a new generation of AI is an important strategic handhold for China to gain the initiative in global science and technology competition … we need to ensure that the core AI technologies are firmly in our own hands" (China 2019).This has clearly demonstrated the mindset of great power competition when introducing China's bold AI initiatives.

Nonetheless, this is only part of the story. Purely focusing on the geopolitical aspect of China's AI plans misses the bigger picture and often assumes a natural top-down translation of these grand words into concrete local practices. Indeed, the local actors' immediate responses to the central state's plans appear to support this top-down observation. In the name of responding to and implementing the central state's AI announcements, over 15 provincial units—including 11 announced between 2017 and 2018—put forward their own local AI policies (Qianzhan 2018). The Shanghai government, for example, introduced its own AI plans, in which it aspires to become a national leading AI city, with some parts of its AI innovation reaching the level of the world leaders by 2020, a world-leading AI centre and an excellent global city by 2030 (Shanghai 2017). Similarly, Jiangsu Province, Beijing City, Zhejiang Province and Liaoning Province set out a series of government policies to foster AI development in their respective regions (Iyiou 2018). The above central-local interaction has created the illusion of a "whole-of-nation" "coordinated"

Chinese approach to AI in that all Chinese state actors selflessly echo in concert the central call for and work to advance China's national interests—a classic top-down national command approach.

Nonetheless, the enthusiastic responses from local states to AI policy slogans such as "to vigorously develop AI" should be critically examined as they are often a political performance, reflecting the art of political correctness (Zeng 2020). In practice, the central-local interaction is far more complicated than a top-down approach. The top-down logic assumes that the central government is the pioneer and leader deciding the strategic direction for China's AI advancement in 2017, followed by local and subnational leaders. However, this may not always be the case. Instead of simply following the central state's instructions, local and subnational actors actively deployed their own AI plans way ahead of the central state's announcement in July 2017 (Ding 2018). The province of Guangdong, for example, had planned its AI policies for years and announced its "2015–2020 Intelligent Manufacturing Development Plan" in February 2016.

Yiou's report shows that in 2015, 11 AI plans were introduced by 10 provincial units, and that in 2016, 9 plans were introduced by 7 provinces (Iyiou 2018). Put another way, a year before the State Council's AI plans were issued in 2017, 14 provinces had already announced 20 AI plans (Iyiou 2018).All of these 14 provinces have introduced additional AI plans since the State Council's AI announcement in 2017. When only observing their policies from 2017 onwards, they do appear to be a response to the central government's mobilization. However, many of their new AI plans from 2017 onwards were a simple continuation and modification of their previous plans. To some extent, some local states were rebranding their existing plans to make those plans look new and a follow-up to the central call. By using this "old wine in a new bottle" tactic, they could gain economic benefits from the central state to conduct their existing plans.

The local states were ahead of not only the State Council's AI paper in 2017 but also almost all Beijing's central agencies' AI plans. The earliest provincial AI plans can be traced back to as early as 2009 with a steady number of plans published afterwards; many were published before 2012 when the central government put forward its Internet of Things policy (CISTP 2018). In this regard, the central actors were neither the pioneers who discovered AI's potential nor the commanders who ordered local actors to develop AI. Driven by the benefits of an AI economy, the local and subnational actors had already introduced their AI agendas in advance of Beijing's call. The central state, at most, formalized the existing grassroots AI initiatives at the national level. In short, the local AI initiatives were already there, and the central government recognized their importance and upgraded them into a national plan. This development suggests more of a bottom-up rather than a top-down manner.

This bottom-up manner is reflected not only in the sequence of policy announcements among state actors but also in the success of private entrepreneurship in China's high-tech industry. Privately owned tech companies—such as Alibaba, Tencent and Baidu—are the key drivers of China's booming internet economy rather than the state instruction from Beijing (Jing 2018). These privately owned internet giants are also the leading forces behind China's AI innovation. China's leading technologies, such as bike-sharing and digital payments, were created neither on Beijing's instruction nor by state funding, but by market entrepreneurship (Jing 2018).These market forces are primarily driven by business profits, not strategic geopolitical interests imposed by a few leaders in Beijing. All of these reflect the success of bottom-up innovation rather than top-down command.

Moreover, the top-down command narrative misguidedly assumes a high level of coordination among domestic actors and a chain of command through a hierarchical bureaucracy to operate the AI economy. The Chinese style of federalism determines that this is not the case. As previously mentioned, China's economy is not centrally planned but heavily reliant on a division of responsibilities between central and local states, giving the latter primary economic responsibility in their jurisdictions. Under this high level of local autonomy and the central government's restricted discretion, it is about not only what the central government wants but also what local states want.

It is true that local governments do share overlapping interests with the central government, i.e. economic growth via boosting the AI economy, which explains why local actors actively jumped on the bandwagon to echo the central call. However, there is also a critical difference in their interests. The central government, with its strategic thinking in terms of making China a leading power, is after the geopolitical interests of China as a whole. The ideal scenario is a concerted national effort to deliver this singular goal, assuming complementary rather than competitive relations among domestic local actors. For local states, however, they are not always driven by the bigger picture of China's national interests but by the desire to maximize their own economic interests. This means that conventional economic competition over factors of production, such as labour and capital, under China's federalism applies in the case of China's AI development.

Indeed, Chinese provinces never try to hide their competitive mindsets. Almost all the Chinese provinces chose a high-profile approach to announce their subsidies policies to boost the AI economy. They put forward similar policies, including competing deals on tax benefits, rent concessions and funding subsidies, and they were fighting for the same AI resources: AI talent, AI investment and AI companies. The shortage in resource supply makes this competition very intense. Estimation shows that, for example, there is a shortfall of over 5 million qualified workers in China's AI industry (Hou et al. 2018); this is hardly surprising given that the shortage of AI talent is a global phenomenon. After 18 years of explosive growth, the total financing of China's AI industry met its first decline in 2019, dropping from 148.453 billion yuan in 2018 to 96.727 billion yuan in 2019—a decrease of 34.8% (CheetahGlobalLab 2020). Following this downward trend, 336 Chinese start-up tech companies shut down their operations in 2019. The year 2019 is widely called "capital winter" in the industry (Yinggu Wang 2019). It remains to be seen how the pandemic has affected China's AI industry.

Many local governments also used the same strategy to build momentum for their AI industries: high-profile AI conferences. This was chosen by many local governments as the first major step after they announced their AI policies. For example, in 2018, more than 8 local provinces or municipalities under the central government including Tianjin, Guizhou, Guangdong, Sichuan, Beijing, Chongqing, Shanghai and Nanjing organized AI conferences. Needless to say, so many conferences in a year kept China's leading internet entrepreneurs, such as Jack Ma, Robin Li and Ma Huateng, very busy. Again, local governments competed to invite these leading figures to boost the influence and profile of their conferences. So, the point to emphasize here is that despite the central planning at the top level, there is a high level of regional competition. As one of my interviewees points out:

> the national planning is relatively important and can help to prevent disorderly and competitive development. However, whether it is implemented at the local level is related to the local governments' ability and awareness. Most of the times, there are irrational competition driven by GDP targets.

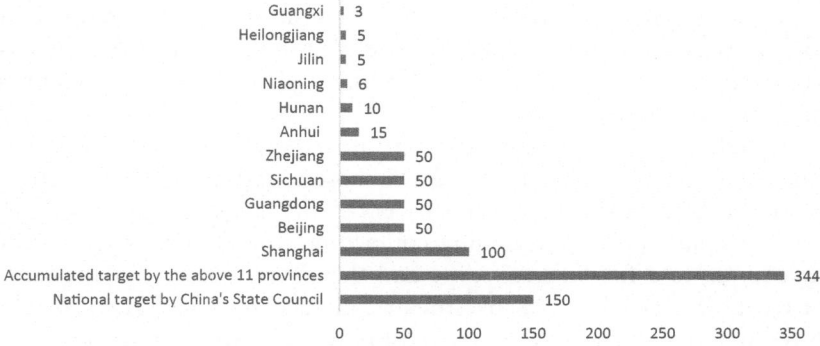

Fig. 2.1 Summary of Chinese provincial-level unit target values for the AI core industry by 2020 (unit: 1 billion yuan)[4]

There are also disorderly competitions like using land and policy resources and financial support to seize the opportunity of AI development.[3]

The disorderly regional competition is further demonstrated by local actors' response to the central government. In 2017, the State Council announced the aforementioned national strategy of AI with the hope of building up a more than 150-billion-yuan AI core-industry and a more than 1000-billion-yuan AI-related industry by 2020. Following this, many provinces also announced their ambitious quantitative targets. Figure 2.1 summarizes their quantitative targets for the AI core industry by 2020.

As Fig. 2.1 shows, the City of Shanghai and Sichuan Province expected to develop AI core industries worth more than 100 and 50 billion yuan by 2020, respectively. Should these goals be achieved, together these two would be sufficient to meet the national target, i.e. the 150-billion-yuan goal set by the State Council. Adding the targets of these 11 provinces

---

[3] Interview with a university-based AI expert in China conducted on 26 March 2020.

[4] This figure is compiled by the author based on public information. Hunan Province's target was for 2021, and Zhejiang Province's was for 2022. A similar statistic conducted by the Qianzhan Industry Research Institute shows that the accumulated target value of the AI core industry for the 12 provinces reaches 429 billion yuan by 2020. See (Qianzhan 2018, 2019).

together, the total value of the AI core industry would be 344 billion yuan—more than double the national target.

This inflation is also reflected in the target value of the AI-related industry. Figure 2.2 summarizes the quantitative targets of the AI-related industry by 2020. As Fig. 2.1 shows, these 10 provincial units' accumulated target value reaches 2,000 billion yuan—twice that of the national target set by the State Council. Notably, the above accumulated provincial target only covers 10–11 provinces, while the State Council's national target was supposed to cover the whole of mainland China, i.e. 31 provincial units. Besides these 10–11 provinces, there were some provincial units that had not put forward their AI plans but were still interested in developing an AI economy, or others who had their AI industry plans but did not specify their quantitative targets for 2020. Had they given a figure, the accumulated provincial target would be even higher.

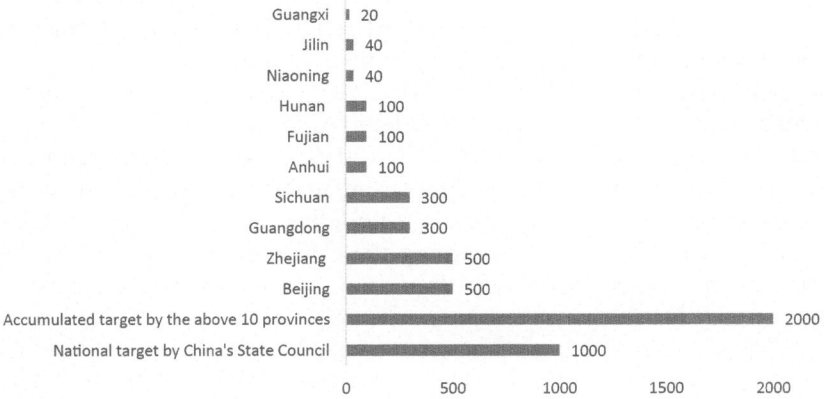

Fig. 2.2 Summary of Chinese provincial-level unit target values of the AI-related industry by 2020 (unit: 1 billion yuan)[5]

[5] This figure is compiled by the author based on public information. Hunan Province's target was for 2021, and Zhejiang Province's was for 2022. A similar statistic conducted by the Qianzhan Industry Research Institute shows that the accumulated target value of the AI-related industry for the 10 provinces reaches 1,715 billion yuan by 2020. See (Qianzhan 2018, 2019).

In addition, the national target of 150 billion yuan is not a moderate target. The widely cited Zhongkegaofu report estimated only a 57-billion-yuan AI core industry in 2019 (Zhao 2019; Zhongkegaofu 2019). Estimations of China's AI industrial value in 2020 ranged from 40 to 160 billion yuan, with most institutes putting forward a number way lower than 150 billion (Qianzhan 2019: Sect. 5–1).[6] In this context, the grossly inflated regional targets are quite unrealistic.[7] Notably, this phenomenon is far from unique. The accumulated regional GDP figures, for example, are always unreliable as China's local actors tend to fake their economic data. The reported local GDP figures often inflate revenues and downplay debts, and the Chinese central government does not always know what is going on (Koch-Weser 2013; Wallace 2016).

Needless to say, this tendency for data manipulation has affected statistics on the AI industry. Indeed, plenty of room was left for this data manipulation. The target numbers in Figs. 2.1 and 2.2 refer to the value of the AI core and the related industry, respectively. However, the line between the AI core industry and the AI-related industry is very blurred, leaving room and flexibility for different interpretations (Ding 2018). The fundamental concept of "AI" is very difficult to define. This umbrella term has made related concepts, including "AI technology", "AI strategy" and "AI industry", very fuzzy. As Ding (2019a) points out, "the concept of AI, which encompasses anything from fuzzy mathematics to drone swarms, becomes so slippery that it is no longer analytically coherent or useful". This lack of clarity regarding these AI concepts and the central government's AI plans has been exploited by some market forces as an opportunity for profits.

While China's AI approach is often labelled as a state-oriented model to be distinct from others such as the US's market-oriented model (Burrows and Mueller-Kaler 2021, 2020; Webb 2019), the role of market forces cannot be underestimated. Some market actors view China's national AI plans as a rare marketing opportunity to hype AI business. They have contributed to a national AI boom carnival in China, leading

---

[6] For other estimations, please see (Kewalramani 2018: 9).

[7] One of my interviewees, however, argues that there is no need to make a fuss over these figures. After all, "unlike poverty alleviation tasks, there is no pass line. The value of AI industry is all up to how you calculate; statistics is nothing more than how you define those rules. The entire AI industry is very vague anyway". Interview with a university-based AI expert in China conducted on 26 March 2020.

to unreasonably high salaries, expectations of investment returns and valuations of unicorns in China's AI industry (Chen 2017). Consequently, there are increasing concerns over an AI bubble in China (Chen 2017; Stars 2019). Although the AI hype is a global phenomenon, the one in China is one of if not the biggest thanks to China's high-profile state AI plans. For some, the current AI bubble in China is "insane" (Stars 2019). Many Chinese AI companies took advantage of AI's fuzzy definition to brand their products with an AI label in order to make more profits and attract more investments (Wang 2018). According to Li Yizeng, Principal Fellow at Shanghai Academy of Social Sciences' Internet Studies Center, 90% of the current AI business products on the Chinese market, such as smart speakers and robots, are not "real" AI technology (Wang 2018). According to Wei Zhe, Founder of Jiayu Fund, 90% of those products are "fake AI" (Wang 2018).This has contributed to a very unhealthy AI market, lacking in genuine innovation.

In addition, some market actors have manipulated the fuzzy definition of an AI company to cheat state funding. According to a Deloitte report, "90% or even 99%" of China's AI companies are fake AI companies (Deloitte 2018). These companies have used their fake AI cover to deceive the government and obtain state subsidies (Deloitte 2018)—a problem that commonly exists in China's other high-tech industries as well. Generally, it exposes the efficiency and quality problems of China's tech policy. While it boosts sector growth by providing more incentives for the private and research sectors and allocating more resources in high-tech industry, it inevitably poses the challenges of over-capacity, wasteful investment and quality innovation. In this regard, China's national AI plan is further distorting the already unhealthy Chinese AI market. When dealing with the flow of state funding, the commercialization of new AI technologies remains a critical challenge for Chinese AI start-ups (Dai 2018). Without finding the real market demand, the current industrial prosperity and growth is neither sustainable nor helpful to its global competitiveness in the long run.

REGIONAL COMPETITION AND CENTRAL COORDINATION: STRATEGIC THINKING VS ECONOMIC DEVELOPMENT

Previous sections have shown that China's AI development suffers from problems of disorderly regional competition, poor national coordination and market manipulation. Why does the central government not intervene to a larger extent and make a nationally concerted AI push? To understand this, it is important to clarify China's goal of AI advancement. While geopolitical thinking has clearly shaped China's AI strategy, the growth of the AI economy is the most important rationale driving China's AI initiatives. Since China's market reforms in the early 1980s, economic growth has become a key pillar of the CCP's legitimacy. Many in China view technological innovation as the key to maintaining China's economic growth as technological advancement will improve overall labour productivity and thus create social wealth (Feng 2018). The intention to boost national wealth through the AI industry is clearly indicated in the aforementioned Chinese State Council's AI plan, in which the quantitative goals of developing a 150-billion-yuan AI industry by 2020, a 400-billion-yuan AI industry by 2025 and a 1000-billion-yuan AI industry by 2030 are clearly set out. Apparently, only when the AI economy has a strong foundation can China deliver Beijing's geopolitical vision of becoming a leading AI powerhouse. In other words, the AI economy is the primary goal, and geopolitical leverage is the add-on.

When it comes to promoting economic growth, a painful lesson that China learnt from Mao Zedong's era is that a centrally planned approach would not work, leading the way for the subsequent pragmatic, market-oriented approach. As previously mentioned, this Chinese style of federalism has been successful in boosting the Chinese economy despite the problems caused, such as declining central authority and regional competition. A higher level of state intervention is usually not good for the market. In fact, the current level of state intervention in AI development has already led to various problems, including the previously mentioned waste of state funding and low efficiency. As long as economic incentives drive China's AI plan, its success relies on market competition and not top-down state intervention.

It may also be worth mentioning that, even for authoritarian regimes like China's, coordination and central planning are not as straightforward as many would expect. Even within the central government in Beijing, bureaucratic politics is everywhere. As far as AI is concerned, jurisdiction among the central state's different departments over China's AI policy is anything but straightforward. Four central agencies, including the National Development Reform Commission, the Ministry of Science and Technology, the Ministry of Industry and Information Technology and the Cyberspace Administration of China, fought to assert their power in deciding and managing China's AI policy (Ding 2018: 15). Different national AI policy papers indicate remarkably interesting conflicts over which agencies have the mandate to command China's AI policy (Ding 2018: 15). Not surprisingly, the Cyberspace Administration of China, China Academy of Information and Communications Technology and Ministry of Science and Technology have been championing three different governance approaches towards AI respectively (Sheehan 2022).

In other words, even central agencies in Beijing are not pursuing a single unified goal—let alone a whole national attempt to advance AI in China. As Ding rightly points out, "although the central government plays an important guiding role, bureaucratic agencies, private companies, academic labs, and subnational governments are all pursuing their own interests to stake out their claims to China's AI dream" (Ding 2018: 15). According to one commenter, for example, the introduction of the State Council's "New Generation AI Development Plan" is China's Ministry of Science and Technology's attempt to assert its power in China's high-tech developments. As the plan establishes an "AI Implementation Office" located in the Ministry of Science and Technology to promote the implementation of AI plans, it gives the Ministry considerable influence in steering China's AI research agenda that was previously driven by other ministries or scientists from the Chinese Academy of Sciences and the Chinese Academy of Engineering (Laskai 2017). Thus, some argue that this additional bureaucratic layer would invite negative impacts on China's AI innovation and technological development (Laskai 2017; Ding 2019b).

Concluding Remarks

As this chapter shows, "to vigorously develop AI" is a broad and vague political slogan to mobilize Chinese domestic actors. Far from being a specific plan, the State Council's "New Generation AI Development Plan" is a "manifesto about the future" (Laskai 2017) or a "wish list" of AI technology that the central state would like to develop with few concrete ideas about how to get it done (Sheehan 2018). Its implementation heavily relies on local and subnational actors to interpret the AI slogan and find their own ways to motivate the private sector and accelerate AI activities within their respective jurisdictions. This process often supports local agendas and interests as the mechanism allows a high level of discretion for local actors to decide local AI activities. This slogan mobilization process means that local and subnational actors play an important role in shaping AI politics.

In this regard, China's AI innovation does not simply follow a top-down command approach, which makes it distinctly different from that in the US and Europe. While strategic thinking and national planning mindsets are clearly there backing the Chinese central state's AI plans, these top-level grand masterplans are not completely unfolded into concrete practices at the local level. The nature of China's economic circumstances means that its AI industry is primarily driven by a range of local, subnational and non-state actors who have diverse—and sometimes competing—interests and little diplomatic and geopolitical awareness. Their struggle for resources has shaped the development of China's AI industry. Instead of a top-down command model, the development of China's AI policies largely follows a bottom-up manner in that existing local AI initiatives successfully won recognition from Beijing and were upgraded to become a national focus.

Rather than a concerted national effort to boost the AI industry, the Chinese approach faces the problems of coordination and manipulation. Similar to the US and Europe, China's market forces and entrepreneurs play a key role in boosting the AI industry, and they are pursuing individual commercial interests not the country's national interests. Notably, this chapter only examines the domestic struggle for AI resources within China. There is also an international dimension. As previously discussed, China's political system has become increasingly internationalized due to globalization. With China's integration into the world, it has been increasingly exposed to foreign influence and shaped by its transnational interests. So is its AI industry. China's AI advancement has largely benefited from international collaboration and access to foreign technology (O'Meara 2019; Hannas and Chang 2019). Many leading Chinese internet companies are partly owned by foreign capital. Not surprisingly, US-China conflicts have led to considerable concerns among Chinese AI scientists and entrepreneurs about their potential negative impact on China's AI innovation. Given that transnational interests play an important role in China's AI innovation, it is misguided to assume highly unified and coherent efforts are being made by China's AI sector to advance Beijing's AI plans and thus China's national interests.

Above all, purely focusing on the great power competition aspect of China's AI development would inevitably neglect the nuanced development on the ground and thus vastly exaggerate (a) Beijing's capacity to mobilize domestic actors to achieve a unified strategic objective and (b) the advancement of China's AI programmes and its threat in geopolitical terms. Thus, two critical reflections are needed. First, as mentioned, many summarize China's approach as a "nationally concerted" and "whole-of-government" approach. As this approach is considered to have a "distinct advantage" in the global AI competition, many call for the US to adopt a similar approach. However, if this characterization of China's AI approach does not reflect the reality, what lesson should be learnt from the Chinese approach?

The second reflection is about how to respond to China's rise as a global AI leader. The exaggerated geopolitical threat of China's AI advancement has (un)intentionally contributed to China's near-peer competitors'—especially the US's and Europe's—unnecessary anxiety. Nowadays, AI innovation has been increasingly influenced by geopolitics. The rise of the AI race narrative has implicitly and explicitly promoted a zero-sum view into the understanding of AI innovation. The narrative's priority on competition over cooperation and destruction over creation has the potential to turn AI in a more nationalistic and inward-looking direction. This focus on geopolitics has contributed to growing attention on the implications of AI for national security, as the next chapter will discuss.

REFERENCES

Breslin, Shaun. 1996. *China in the 1980s: Centre-Province Relations in a Reforming Socialist State*. Basingstoke: Palgrave Macmillan.

Burrows, Mathew, and Julian Mueller-Kaler. 2021. *Smart Partnerships amid Great Power Competition: AI, China, and the Global Quest for Digital Sovereignty*. The Atlantic Council GeoTech Center.

Callahan, William. 2016. *China's Belt and Road Initiative and the New Eurasian Order*. Norwegian Institute of International Affairs (Oslo).

CheetahGlobalLab. 2020. 泡沫挤压: *AI行业热度骤降, 基础层投资被忽视 (Bubble Squeeze: AI Industry Plummets and Infrastructure Investment Is Ignored)*. 猎豹全球智库 (Cheetah Global Lab). Available at https://36kr. com/p/5282642. Accessed on 29 Feburary 2020.

Chen, Celia. 2017. "China's Artificial Intelligence Sector in Danger of Becoming a 'Bubble', Experts Warn." *South China Morning Post*. Available at https://www.scmp.com/tech/innovation/article/2082217/chinas-artificial-intelligence-sector-danger-becoming-bubble-experts. Accessed on 29 Feburary 2020.

China. 2015. "国务院关于印发促进大数据发展行动纲要的通知 (State Council's Decision on Promoting the Development of Big Data)." Available at http://www.gov.cn/zhengce/content/2015-09/05/content_10137.htm. Accessed on 20 November 2015.

———. 2019. 人工智能具有很强的"头雁"效应 (*Artificial Intelligence Has a Strong "Head Goose" Effect*). Available at http://paper.people.com.cn/rmr bhwb/html/2019-07/26/content_1938122.htm. Accessed on 28 Feburary 2020.

CISTP. 2018. 中国人工智能发展报告2018 (*2018 Report on China's AI Development*). Beijing: China Institute for Science and Technology Policy at Tsinghua University (CISTP).

Corrigan, Jack. 2018. "U.S. Needs a National Strategy for Artificial Intelligence, Lawmakers and Experts Say." *Defense One*.

Dai, Sarah. 2018. "Investor warns of Day of Reckoning for 90 per cent of Chinese AI Start-Ups as Funding Dries Up." *South China Morning Post*.

Deloitte. 2018. 中国人工智能产业白皮书 (*White Paper on China's AI Industry*). Deloitte.

Ding, Jeffrey. 2018. *Deciphering China's AI Dream*. Future of Humanity Institute, University of Oxford. Available at https://www.fhi.ox.ac.uk/deciph ering-chinas-ai-dream/. Accessed on 28 Feburary 2020.

———. 2019a. *"China's Current Capabilities, Policies, and Industrial Ecosystem in AI" Testimony before the U.S.-China Economic and Security Review Commission Hearing on Technology, Trade, and Military-Civil Fusion: China's Pursuit of Artificial Intelligence, New Materials, and New Energy*.

———. 2019b. "The Interests Behind China's Artificial Intelligence Dream." In *Artificial Intelligence, China, Russia, and the Global Order: Technological, Political, Global, and Creative Perspectives*, edited by Nicholas Wright, 43–47. Air University Press.

Feng, Shuai. 2018. "人工智能时代的国际关系:走向变革且不平等的世界 (International Relations in the Age of AI: Moving Towards Shifting and Unequal World)." 外交评论 (*Foreign Affairs Review*) 1: 128–156.

International Crisis Group. 2012. "Stirring Up the South China Sea (I)." Asia Report No. 223.

Hameiri, Shahar, and Lee Jones. 2016. "Rising Powers and State Transformation: The Case of China." *European Journal of International Relations* 22 (1): 72–98.

Hameiri, Shahar, and Jinghan Zeng. 2020. "State Transformation and China's Engagement in Global Governance: The Case of Nuclear Technologies." *The Pacific Review* 33 (6): 900–930.

Hameiri, Shahar, Lee Jones, and John Heathershaw. 2019. "Reframing the Rising Powers Debate: State Transformation and Foreign Policy." *Third World Quarterly* 40 (8): 1397–1414.

Hannas, Wm. C., and Huey-meei Chang. 2019. *China's Access to Foreign AI Technology: An Assessment.* Center for Security and Emerging Technology.

Heilmann, Sebastian. 2008a. "From Local Experiments to National Policy: The Origins of China's Distinctive Policy Process." *The China Journal* 59: 1–30.

———. 2008b. "Policy Experimentation in China's Economic Rise." *Studies in Comparative International Development* 43: 1–26.

———. 2009. "Maximum Tinkering Under Uncertainty, Unorthodox Lessons from China." *Modern China* 35 (4): 450–462.

———. 2018. *Red Swan: How Unorthodox Policy-Making Facilitated China's Rise.* Hong Kong: The Chinese University of Hong Kong Press.

Hill, Christopher. 2016. *Foreign Policy in the Twenty-First Century.* Basingstoke: Palgrave Macmillan.

Hou, Liqiang, Chenglong Jiang, and Fangjie Zhu. 2018. "Competition for Talent Intensifies as China's AI Industry Develops." *China Daily*. Available at http://europe.chinadaily.com.cn/a/201802/05/WS5a77b4aca3106e7dcc13aba2.html. Accessed on 29 Feburary 2020.

Ives, Jaqueline, and Anna Holzmann. 2018. *Local Governments Power Up to Advance China's National AI Agenda.* Mercator Institute for China Studies. Available at https://www.merics.org/en/blog/local-governments-power-advance-chinas-national-ai-agenda. Accessed on 28 Feburary 2020.

Iyiou. 2018. *19省、市人工智能相关政策研究(上) (Studies on the Relevant AI Policies of 19 Provinces).* Iyiou. Available at https://www.iyiou.com/intelligence/insight65899.html. Accessed on 29 Feburary 2020.

Jing, Meng. 2018. *Is Xi Jinping's Iron Grip Better Than Adam Smith's Invisible Hand for Technology Innovation?* Available at https://www.scmp.com/tech/article/2173128/xi-jinpings-iron-grip-better-adam-smiths-invisible-hand-technology-innovation. Accessed on 28 Feburary 2020.

Jones, Lee. 2019. "Theorizing Foreign and Security Policy in an Era of State Transformation: A New Framework and Case Study of China." *Journal of Global Security Studies* 4 (4): 579–597.

Jones, Lee, and Jinghan Zeng. 2019. "Understanding China's 'Belt and Road Initiative': Beyond 'Grand Strategy' to a State Transformation Analysis." *Third World Quarterly* 40 (8): 1415–1439.

Jones, Lee, and Yizheng Zou. 2017. "Rethinking the Role of State-owned Enterprises in China's Rise." *New Political Economy* 22 (6): 743–760.

Kewalramani, Manoj. 2018. *China's Quest for AI Leadership: Prospects and Challenges.* Takshashila Institution.

Koch-Weser, Iacob. 2013. *The Reliability of China's Economic Data: An Analysis of National Output.* The U.S.-China

Laskai, Lorand. 2017. "Beijing's AI Strategy: Old-School Central Planning with a Futuristic Twist." *Council on Foreign Relations*.

Leverett, Flynt, and Bingbing Wu. 2017. "The New Silk Road and China's Evolving Grand Strategy." *The China Journal* 77 (1): 110–132.

Lieberthal, Kenneth. 1992. "Introduction: The 'Fragmented Authoritarianism' Model and its Limitations." In *Bureaucracy, Politics and Decision Making in Post-Mao China*, edited by Kenneth Lieberthal and David Lampton, 1–31. Berkeley and London: University of California Press.

Lieberthal, Kenneth, and David Lampton. 1992. *Bureaucracy, Politics, and Decision Making in Post-Mao China*. University of California Press.

Lieberthal, Kenneth, and Michel Oksenberg. 1988. *Policy making in China*. Princeton University Press.

Miller, Tom. 2017. *China's Asian Dream: Quiet Empire Building Along the New Silk Road*. London: Zed Books.

Montinola, Gabriella, Yingyi Qian, and Barry Weingast. 1996. "Federalism, Chinese Style: The Political Basis for Economic Success." *World Politics* 48 (1): 50–81.

Mueller-Kaler, Julian. 2020. *Europe's Third Way*. Atlantic Council. Available at https://www.atlanticcouncil.org/content-series/smart-partnerships/europes-third-way/. Accessed on 16 September 2021.

O'Meara, Sarah. 2019. "AI researchers in China Want to Keep the Global-Sharing Culture Alive." *Nature* 569.

Parris, Kristen. 1993. "Local Initative and National Reform: The Wenzhou Model of Development." *The China Quarterly* 134: 242–263.

Qian, Yingyi, and Barry Weingast. 1995. "China's Transition to Markets: Market-Preserving Federalism, Chinese Style." *Journal of Policy Reform* (2): 149–185.

Qianzhan. 2018. 一文带你了解2018年全国各地人工智能行业最新政策! *(One Article to Help You to Know Latest Regional AI Policies)*. Qianzhan Industry Institute. Available at https://www.qianzhan.com/analyst/detail/220/180329-8cef9d2f.html. Accessed on 29 Feburary 2020.

———. 2019. 2019年人工智能行业现状与发展趋势报告 *(2019 Report on Industry Status and Development Trends of AI)*. Qianzhan Institute. Available at https://bg.qianzhan.com/report/detail/1910081709070618.html. Accessed on 29 Feburary 2020.

Schurmann, Franz. 1966. *Ideology and Organization in Communist China*. Berkeley and Los Angeles: University of California Press.

Shanghai. 2017. 关于本市推动新一代人工智能发展的实施意见 *(Opinions on the Implementation of this City's New Generation of AI Development)*. Shanghai Government. Available at http://www.shanghai.gov.cn/nw2/nw2314/nw2319/nw12344/u26aw54186.html. Accessed on 29 Feburary 2020.

Sheehan, Matt. 2018. "How China's Massive AI Plan Actually Works." *Macro Polo*.

Sheehan, Matt. 2022. *"China's New AI Governance Initiatives Shouldn't Be Ignored"*. Carnegie Endowment for International Peace. Available at https://carnegieendowment.org/2022/01/04/china-s-new-ai-gov ernance-initiatives-shouldn-t-be-ignored-pub-86127. Accessed on 14 March 2022.

Stars. 2019. *The Chinese Bubble in Artificial Intelligence Is Insane.* Stars Insights. Available at https://www.the-stars.ch/wp-content/uploads/2019/07/HUANG-Yuanpu_The-Chinese-Bubble-in-Artificial-Intelligence-is-Ins ane.pdf. Accessed on 29 Feburary 2020.

Su, Xiaobo. 2015. "Nontraditional Security and China's Transnational Narcotics Control in Northern Laos and Myanmar." *Political Geography* 48: 72–82.

Wallace, Jeremy. 2016. "Juking the Stats? Authoritarian Information Problems in China." *British Journal of Political Science* 46 (1): 11–29.

Wang, Shaoguang. 2009. "学习机制，适应能力与中国模式[Learning Mechanism, Adaptability, and Chinese Model]." 开放时代 *(Open Times)* (7).

Wang, Xinxi. 2018. 伪概念泛滥，认知模糊，AI营销或需一场再教育 *(Pseudo-Concepts Are Proliferated and Cognition Is Fuzzy, AI Marketing May Need a Re-education)*. Iyiou. Available at https://www.iyiou.com/p/80682.html. Accessed on 29 Feburary 2020.

Wang, Yinggu. 2019. 面对技术与市场"双瓶颈"，中国人工智能企业准备迎接寒冬 *(Facing the "Double Bottlenecks" of Technology and Market, Chinese AI Enterprises Are Preparing for the Cold Winter)*. China Money Network. Available at https://www.zhongguojinrongtouziwang.com/2019/03/07/70351/. Accessed on 29 Feburary 2020.

Webb. 2019. *The Big Nine: How the Tech Titans and Their Thinking Machines Could Warp Humanity.* PublicAffairs.

Xi, Jinping. 2017. *The Chinese Communist Party's 19th Party Congress Report.*

Xing, Yijun, Yipeng Liu, and Cooper Cary. 2018. "Local Government as Institutional Entrepreneur: Public–Private Collaborative Partnerships in Fostering Regional Entrepreneurship." *British Journal of Management* 29 (4): 670–690.

Xu, Chenggang. 2011. "The Fundamental Institutions of China's Reforms and Development." *Journal of Economic Literature* 49 (4): 1076–1151.

Yu, Yifan. 2019. "Why China's AI Players Are Struggling to Evolve Beyond surveillance." *Nikkei Asian Review.* Available at https://asia.nikkei.com/Spotlight/Cover-Story/Why-China-s-AI-players-are-struggling-to-evolve-bey ond-surveillance. Accessed on 1 March 2020.

Zeng, Jinghan. 2015a. "Did Policy Experimentation in China Always Seek For Efficiency?—The Case of Wenzhou Financial Reform." *Journal of Contemporary China* 24 (92).

———. 2015b. *The Chinese Communist Party's Capacity to Rule: Ideology, Legitimacy and Party Cohesion.* Palgrave Macmillan.

Zeng, Jinghan. 2019. "Narrating China's Belt and Road Initiative." *Global Policy* 10 (2): 207–216.

Zeng, Jinghan. 2020. *Slogan Politics: Understanding Chinese Foreign Policy Concepts*. London: Palgrave Macmillan.

Zhao, Yuhan. 2019. 人工智能核心产业规模达570亿元 (Value of AI Core Industry Reaches 57 Billion Yuan). *Beijing Daily*. Available at http://bjrb. bjd.com.cn/html/2019-08/12/content_12059007.htm. Accessed on 29 Feburary 2020.

Zheng, Yongnian. 2007. *De Facto Federalism in China: Reforms and Dynamics of Central-Local Relations*. Singapore: World Scientifics.

Zhongkegaofu. 2019. *2019年中国人工智能产业研究报告 (2019 Report on China's AI Industry)*. 中科高服 (Zhongkegaofu). Available at https://mp. weixin.qq.com/s/BGBZ-aqd_AMeuFUYbnAjFg. Accessed on 29 Feburary 2020.

CHAPTER 3

China's Security Politics of AI

INTRODUCTION

Driven by the focus on geopolitical implications of AI development, national states have been paying close attention to its impact on national security. The military application of AI technology, for example, has led to a discussion about a global AI race (Merz 2019; Kempe 2019; Castro et al. 2019). As previous chapters discussed, as a key global AI player, China has made an ambitious three-step plan to become a leading AI power by 2030 (China 2017). Many international analysts characterize this Chinese AI approach as a "unified/integrated" and "nationally concerted" effort (Liu 2019; Ramanathan 2019; Ives and Holzmann 2018; Jing 2018; Yu 2019; Lanier and Weyl 2020; Lee 2018). This "whole-of-nation/government/society" Chinese approach is argued to have a "distinct advantage" over the US's (Hoadley and Sayler 2020: 24). The relevant arguments explicitly and implicitly indicate a geopolitically driven Chinese AI plan, winning the global AI race and threatening American AI supremacy and national security. In this context, AI is no longer a normal technology but a matter of national and international security.

A similar trend towards AI has taken place in China. In 2016/2017, the Chinese government announced AI as a strategic industry and officially adopted a national approach to boost the AI industry (China 2017). This chapter argues that the Chinese central government is securitizing AI

J. Zeng, *Artificial Intelligence with Chinese Characteristics*, https://doi.org/10.1007/978-981-19-0722-7_3

in order to advance its AI agenda. Its AI policy discourse describes security as one of the most important policy goals. As the State Council of China's "New Generation AI Development Plan" states:

> The world's major developed countries are taking the development of AI as a major strategy to enhance national competitiveness and ***protect national security***... At present, China's situation in ***national security*** and international competition is more complex, and [China] must ... firmly seize the strategic initiative in the new stage of international competition in AI development, to ... effectively ***protecting national security***. (Webster et al. 2017b: emphasis added)

By using securitization as an analytical framework, this chapter examines Chinese AI discourse. As part of the national AI campaign to mobilize Chinese society, to label AI as a security matter is one of the political tactics to gain domestic support. As this chapter will show, national security underlies China's overall strategic thinking of AI with specific reference to its military application and practical use to protect regime security.

This chapter also explores how the Chinese AI approach and its security logic is embedded in China's historical, geopolitical and domestic contexts. China's desire for AI sits in the wider context of its pursuit of modern technologies, and it is driven by China's strong anxiety of technology competition generated from its historical discourse, which blames China's "century of humiliation" on its failure in previous global technology competitions. In this regard, historical experience of "humiliation", including not only being militarily invaded but also falling behind in tech development, justifies the need for contemporary mobilization to avoid repeating history. Geopolitically speaking, China's national approach to AI and the move to make it a security matter is accelerated by increasingly competitive US-China relations. Both sides label the other's AI advancement as a threat and thus accelerate the securitization process. In the domestic arena, with regime security of the CCP as the primary concern, the practical use of AI and its relevant discourses are geared towards the goal of securing the authoritarian rule. In this regard, China's bold AI experiments are practising a unique digital technocracy, making China's AI approach distinct from that in Western societies, as Chapter 4 will discuss.

Currently, this securitization is an ongoing process. Although it remains unclear to what extent the targeted audiences—including local governments, market actors, intellectuals and the general public—are impacted by securitization, they have enthusiastically echoed the central government's AI campaign. In this regard, this securitization certainly helps to convince domestic actors. However, it also brings about unintended consequences including (a) to make China less attractive to global AI labour and capital by producing a nationalistic environment, (b) to hinder industrial efficiency by focusing on self-reliance, (c) to make it harder for China to lead global AI governance, (d) to further reinforce technological rivalry by neglecting the potential of global AI cooperation and (e) to constrain Chinese AI companies' global access. All of these could undermine China's key objectives of fostering a booming AI economy and becoming a global AI leader.

Conceptual Framework: Securitization

This chapter adopts the concept of securitization from critical security studies to analyse China's security politics and AI. Securitization is primarily associated with the "Copenhagen School". It is developed from the works of the School's leading scholars including Barry Buzan and Ole Wæver that consider security a "speech act"(Buzan et al. 1998; Buzan 1983)—"by saying, something is done" (Nyman 2018). Securitization refers to the discursive process in which actors—usually elites and state actors—transform a particular issue into a security matter. During this process, actors will label a particular issue as a security threat and list it as part of a security agenda to justify extraordinary countermeasures. Securitization is considered successful if the relevant audience accepts that the issue in question is a security threat, enabling emergency measures.

This securitization process involves a series of key terms including "securitizing actor", "securitizing move", "referent object" and "audience" (Nyman 2018). Securitizing actor refers to the person or actor who labels a matter as a security issue, and this attempt is a securitizing move. Referent object refers to what is labelled as an object that needs to be protected from the claimed security threat during the securitizing move. Audience is the group for whom this securitizing move performs and who needs to be convinced so that extraordinary measures to deal with the security threat can be accepted.

Table 3.1 Using securitization as an analytical framework to study AI in China

Securitizing actor	*The central government of China*
Securitizing move	Labelling AI advancement as a security matter to mobilize domestic actors
Referent object	The Chinese nation *The Chinese nation and its national interests need to be protected*
Audience	Chinese local governments + market actors + intellectuals + the general public
Facilitating conditions	Historical anxiety about falling behind during global technology competition + geopolitical competition + non-traditional security needs

Table 3.1 categorizes the arguments of the chapter into the securitization framework.[1] It argues that the Chinese central government as the

[1] Originated from the Western-centric Copenhagen School, most studies on securitization are based in a Western—or perhaps more accurately European—democratic context. After all, securitization is a process where an actor moves a particular matter out of the "normal" state of affairs into an emergency national security agenda (Buzan et al. 1998). This usually requires a liberal democratic society to represent a regular democratic politics that emergency national security politics can emerge from. It, however, inevitably makes securitization theory less useful to understanding security in a non-liberal-democratic political order and thus not only limits the scope of the securitization agenda but also fails to realize *international* relations. In the context of global power transition in which Western dominance is in decline, many including Buzan are calling for a more global understanding of international relations and securitization (Kapur and Mabon 2018; Acharya and Buzan 2007; Vuori 2008). This requires considerable attention on security politics in non-Western societies. Needless to say, this admirable non-Western politics agenda faces critical challenges as security politics operates in a very different way within a non-democratic—and often authoritarian and illiberal—setting. Despite so, early attempts have been made to apply securitization theory to explain security dynamics from the Middle East including Egypt (Greenwood and Wæver 2013), Saudi Arabia and Bahrain (Mabon 2018), Africa (Ezeokafor and Kaunert 2018), Latin America including Brazil and Mexico (Gledhill 2018), North America including Cuba (Holbraad and Pedersen 2012), South Asia including India (Kapur 2018) to Central Asia including Kyrgyzstan (Wilkinson 2007). In the case of China, securitization has been proven useful to understanding climate and energy politics (Nyman and Zeng 2016; Bo 2016; Trombetta 2019). Vuori's research, for example, shows its explanatory power in studying political crises in the eras of Mao Zedong, Deng Xiaoping and Jiang Zemin, suggesting that in the Chinese context the principal audience of the securitizing move is not the general public but elites who have the power to shape the security agenda (Vuori 2008, 2011). This chapter focuses on how AI is securitized within China. Needless to say, the securitization process works differently in the Chinese context given its unique state-society relations. Despite a non-liberal

securitizing actor is performing a *securitizing move* by labelling China's AI advancement as a matter of security. In the relevant discourses, the national interests and survival of the Chinese nation are the *referent object* that needs to be protected. As part of the central government's AI campaign to mobilize domestic actors, this performative act aims to convince the domestic *audience* including local, subnational, academic actors, market actors and the mass, as the rest of the chapter will explore.

LABELLING AI AS A SECURITY MATTER

In China, AI is not only a buzzword but also a popular policy slogan. In order to boost its AI industry, the Chinese central government has announced a series of AI policies, as discussed in previous chapters. For example, in May 2016, the National Development and Reform Commission, the Ministry of Science and Technology, the Ministry of Industry and Information Technology and the then Central Leading Group for Cyberspace Affairs jointly released the "'Internet + ' AI Three Years Implementation Plan" (China 2016a). By then, this document mentioned "security" 16 times including 5 references to cybersecurity with none about national security.

A year later, in July 2017, an overall national security approach became evident when the "New Generation AI Development Plan" was put forward by the Chinese State Council. In this authoritative document (China 2017), the word "security" appears 48 times including 8 references to "national security". As the quote in the introduction section of this chapter shows, this document explicitly claims AI as a matter of China's national security. It also clearly points out that

> [China] must take the initiative to ... lead the world in new trends in the development of AI, serve *economic and social development*, and support *national security*. (Webster et al. 2017b: emphasis added)

This indicates that boosting economic growth and protecting national security are the two most important overall goals, as will be discussed later. This highlights the critical role of national security in China's strategic thinking of AI.

democratic setting, there is still a need to convince the domestic audience and thus win more support for China's national AI plan.

As discussed in Chapter 2, before the central government released its AI plan around 2016/2017, some provincial and municipal governments had already developed their own regional policies to boost the AI economy. Many regional AI policies were announced even before the central government introduced the Internet of Things policy in 2012, some of which could be traced back to as early as 2009 (CISTP 2018). At the time, AI was far from a security matter.[2] Scroll forwards a few years to 2016, however, a U-turn of securitizing AI became evident, as indicated in the State Council's AI plan.

The relevant securitizing move belongs to a type of securitization referring to a directive elementary speech that is performed to raise an item on the agenda (Vuori 2008). It consists of "three sequential, elementary speech acts" including claim, warn and request (Vuori 2008: 80). In this case, the State Council of China aims to raise the awareness of its audience about AI's importance and requests the relevant actions. As the quote in the introduction section of the chapter *claims* and *warns*, other countries (i.e. China's competitors) are elevating AI as a significant national strategy for the sake of national security.

The document also *claims* that China's overall AI development is already behind other great powers as China lacks significant original AI innovations (China 2017). As such, it *requests* the nation to prioritize AI advancement to protect national security. This request is followed by setting not only a broad goal of making China a leading AI power but also a three-step plan and a targeted timeline: (1) to catch up with the AI technological progress of world-leading countries such as the US by 2020, (2) to make major breakthroughs in some AI technologies by 2025 and (3) to become a global leading AI power by 2030 (China 2017).In short, the Chinese central government is labelling AI as a national security matter and highlighting the threat of falling behind in order to convince domestic actors to support its planned actions.

Within this overall strategic thinking, a more explicit security position can be found in the official discourse about AI's military application. The State Council's AI plan mentions "national defense" 11 times and states that AI will be able to "elevate national defense strength and assure and protect national security" (Webster et al. 2017b). This plan considers

---

[2] Most of those regional policies were made for boosting the local AI economy, while overall national security was not the focus.

strengthening military-civilian integration as one of six tasks, and states that

> [China shall] deepen implementation of military-civilian integration development strategy, to promote the formation of an all-element, multi-field, high efficiency AI military-civilian integration pattern… strengthen a new generation of AI technology as a strong support to command and decision-making, military deduction, defense equipment, and other applications…promote all kinds of AI technology to become quickly embedded in the field of national defense innovation. (Webster et al. 2017b)

This fits in China's broader military-civilian integration efforts that have become a national strategy since 2015. To put this national strategy into practice, a Central Commission for the Development of Military-Civilian Integration was created in 2017 and headed by the Chinese President Xi Jinping. Given its importance, it is hardly surprising that military-civilian integration applies the use of AI. Indeed, early signs of this approach can be found in the 2015 Chinese Defense White Paper on Strategy. The paper noted the critical importance of the development of intelligent weapons and implications for China's military security (China 2015).

Historical Context: AI Supremacy for National Survival?

National security and economic growth are considered by the State Council as two overall goals of China's ambitious AI plan, as previously mentioned. Given that economic growth represents the most important source of political legitimacy in China (Zhao 2009; Wang 2005b; Laliberté and Lanteigne 2008; Shambaugh 2001; Perry 2008; Krugman 2013; Wang 2005a), it is quite understandable that China's AI plan is pursuing a booming AI economy. However, why is national security listed as an equally important goal? Why does security occupy such a supreme place in China's overall strategic thinking of AI? This should be understood in the wider historical context of China's pursuit of cutting-edge technology.

Although China's AI plans gradually came to public attention on the global stage during the period 2016/2017, they have long historical roots and their development is clearly path-dependent. All those plans were broadly consistent with China's 13th Five-Year Plan and the state-driven industrial plan "Made in China 2025" released in 2015. For example, the concept of AI appeared in the State Council's 13th Five-Year Plan of China in March 2016 along with 5G, big data and cloud computing, which were also considered national priorities (China 2016b). As such, the advancement of AI is part of the tech package to develop China as a leading technological power.

China's ambitious AI strategy and more broadly its technological aspiration is heavily shaped by Chinese discourse of its modern history. China has always had a high level of anxiety about lagging behind *again* in the game of global technology competition. This Chinese obsession with technology is relevant to its discourse about the relations between China and previous industrial revolutions (Zhang 2012; Li 2016; Jin 2019).[3] While it is a common practice to discuss AI in the discourses of industrial revolutions in many international analyses, the Chinese angle of industrial revolutions is critical to understanding its technological aspiration.

In the Chinese official discourse, China as the Middle Kingdom had been the leading superpower until the nineteenth century when it was first defeated by a Western power, Britain. To many in China, it was the first industrial revolution that—beginning in Britain in the eighteenth century represented by the creation of the steam engine—made Britain a leading power (Jin 2019; Wang 2016). It gave Britain the military might to defeat the Qing Dynasty—the then Asian if not global ruler (Jin 2019). It is argued that the Qing Dynasty failed to catch up with the technological innovation, and this failure marked the starting point of China's decline and the so-called "century of humiliation" (Zhang 2012).

This defeat further put China in an unfavourable position during the second industrial revolution (Li 2016). When the revolution took place in the early twentieth century, China was still in the transition from the last feudal dynasty (i.e. Qing Dynasty) to a republic (i.e. Republic of China). The high level of political and social turbulence let China miss the great opportunity to develop itself (Li 2016). When it came to the third industrial revolution launched by the development of the internet and computers, starting around the 1940s/1950s, the People's Republic of China was just founded with serious domestic and international turmoil.

---

[3] This industrial revolution discourse is also evident in the official policy documents. As the Chinese State Council's 2017 "New Generation AI Development Plan" points out, "AI has become the core driving force for a new round of industrial transformation, [which] will advance the release of the huge energy stored from the previous scientific and technological revolution and industrial transformation" (Webster et al. 2017b).

In the end, China missed its chance again—although some argue that it caught the second half (Li 2016). As summarized by Jin Canrong, a leading Chinese public intellectual and policy advisor:

Obviously, all those three industrial revolutions have one common feature – all are made in the West and its consequence is to let the West stay ahead of productivity. On the contrary, China failed to grasp any of those three industrial revolutions and thus stays behind of productivity. As a result, despite years of efforts, China remains a developing country. (Jin 2019)

The relevant historical discourses about technological revolutions are explicitly indicated in Chinese official documents and top leaders' remarks. As Xi Jinping points out during a national conference on science and technology:

historical experience shows that scientific and technological revolutions can always profoundly change the world's development landscapes... some countries have seized the rare opportunity of scientific and technological revolutions to achieve a rapid growth in economic strength, scientific and technological strength, and national defence strength, and a rapid increase in overall national strength... during over 5,000 years of civilization development, the Chinese nation has achieved world-renowned scientific and technological achievements... (however), since modern times, due to various reasons at home and abroad, our country has repeatedly missed the opportunity of scientific and technological revolutions and fallen from a world power to a semi-colonial and semi-feudal country that was bullied by others. Our nation has experienced more than a century of aggression by foreign powers, endless wars, social turmoil, and people's displacement. (Xi 2016)

To many in China including Xi Jinping, the rise of new technologies such as AI, Internet of Things, cloud computing, big data, new energy and 3D marked the beginning of the fourth industrial revolution (Li 2016; Wang 2016; Xi 2016).Unlike previous industrial revolutions in which China was not economically resourced and lacked a favourable socio-political environment, China is now eager not only to jump into but also to lead this fourth revolution (Li 2016). As Jin Canrong elaborates,

the fourth industrial revolution is the biggest historical opportunity for China. Logically, if China grasps this opportunity, in the future, the best technology and industry of humanity will be in China. So, we must grasp this opportunity. (Jin 2019)

In short, the lesson that China learnt from its modern history is that China must master the leading technology for the sake of its national survival, and that this wave of technological development is the "train" that China cannot afford to miss (Zhang 2012; Jin 2019). Although it is debatable whether those discourses are an accurate reflection of the history, they heavily influence the Chinese quest for technological leadership.

Needless to say, that behind these historical discourses and Chinese technological aspirations is China's quest for national rejuvenation. It is the desire that China can take advantage of this wave of technological revolution led by digital technology including AI to return to its "rightful" place, i.e. the superpower status before being defeated by Britain in the nineteenth century. In this regard, this technological ambition will inevitably clash with American supremacy and sit firmly in the arena of great power politics, as the following section will explore.

Geopolitical Context: Competing for AI Supremacy?

Because of the above historical context, China feels a high level of necessity to master key technology; and reliance on foreign technology is considered a "security risk" (Webster et al. 2017a). This Chinese risk awareness has been further strengthened by its conflicts with the US. China's technological aspirations such as the aforementioned "Made in China 2025" package including AI are all considered by many in the US as a security threat, leading to considerable tension between these two countries. By cutting China's tech companies such as Huawei and ZTE off from global semiconductor suppliers (Zhong 2020), the Trump administration had caused existential crises to those Chinese companies. The US's sanctions also targeted Chinese AI start-ups and restricted the export of American AI software to China, hoping to slow down their development (Xu and Naomi 2019). In the meanwhile, the US has made unilateral efforts to pressure its European partners to prevent the flow of advanced technology to China. For example, by appealing on the grounds of security interests, the Trump administration launched a lobby campaign to block the sale of Dutch manufactured computer chip-making machines to China (Alper et al. 2020).

Severe damage caused by American sanctions on Chinese tech companies not only reminds China about its technological weakness but also strengthens its feeling of insecurity about any global reliance. Understandably, China wants to master leading AI technology by itself. As Xi Jinping elaborates,

> accelerating the development of a new generation of AI is an important strategic handhold for China to gain the initiative in global science and technology competition...We need to ensure that the core AI technologies are firmly in our own hands. (China 2019)

China's AI aspiration is also about global leadership. With its rise, China is not fully satisfied with the US-led global order as it feels that it does not have enough say in global norms and rules (Zeng and Breslin 2016). Instead of a norm-taker, China now aspires to be a norm-shaper or even a norm-maker. Many Chinese scholars argue that the current established norms are primarily serving the interests of others not China (Zeng and Breslin 2016). In order to maximize Chinese interests, future norms should be defined by/for China and on Chinese terms.

In reality, such changes can hardly take place in traditional fields where rules are established and changes are likely to lead to resistance. However, AI, as an emerging field, is relatively blank, where rules and norms are waiting to be written. China is now prepared to fill the gap. According to the State Council's AI plan, China would initially establish technical standards of AI by 2020 and will promote the establishment of international AI organizations (China 2017). In January 2018, during the AI Standardization Forum, a Chinese White Paper on AI Standardization was published to advance a framework of AI standards (NISSTC 2019). In April of the same year, Beijing hosted the inaugural plenary meeting—ISO/IEC JTC 1/SC 42—of the international standards committee responsible for setting up international AI standards. Fu Ying, a leading Chinese diplomat, also expressed China's interest in taking the lead to define norms to mitigate the risk of AI's military use (Allen 2019). In the meanwhile, Chinese scholars have also widely discussed future AI governance and how China can play a leadership role (Gao 2017; Feng 2018). All of these indicate a Chinese will to lead global AI governance.

When it comes to global AI competition, many in China consider it "a race of two giants" between China and the US (Allen 2019). According to the Director of AI and the Big Data Index Institute at East China University of Political Science and Law Gao Qiqi, for example, the global order has been shifting from US-led unipolarity towards multipolarity; however, AI development may reverse this trend given American "super-power status" in the AI field (Gao 2017). Gao argues that the gap between the US and others has been widening in the AI field, while the gap between the US and China is shrinking (Gao 2017).In this sense, AI competition is between China and the US, and its outcome will have significant implications on global order (Gao 2017).[4]

Geopolitical competition of AI is fiercer when it comes to AI's military use, where it is directly linked with national security. As previously discussed, the State Council's AI plan has clearly indicated a strong Chinese will to enable its army through AI. China's defence sector has heavily invested in leading AI technology, such as swarming, robotics and machine learning, in the hope of making its weapons systems "intelligentized" and thus develop a "world class" and "modernized" army (Kania 2020). This trend has been closely followed by many international analyses that examine (1) to what extent AI-enabled technology will advance China's military capacity, (2) to what extent this AI-empowered Chinese army will export its technologies to other countries, especially US adversaries and (3) what the US should do as a countermeasure in light of the security threat. Chinese analysts are also closely observing the global trend of AI's military practices and thus advise how the Chinese army can not only learn from others' practices but also cope with this external security threat (Lei et al. 2019). For example, the Chinese defence community closely followed the development of the US's "Third Offset" strategy in 2014 and reacted by reflecting its own military modernization approach with reference to AI technologies (Wood 2016). In these regards, AI's military use does carry considerable strategic risks of enforcing securitization and thus the likelihood of conflicts.

As such, both China and the US are using each other as a mirror to reflect what they should and could do. In other words, both sides have been labelling the other as a security threat to justify their preferred AI agenda. In addition to the aforementioned example of the US's "Third

---

[4] Similarly, Zhu Min, a Professor at Tsinghua University and the former Deputy Managing Director of the International Monetary Fund, argues that, given the overwhelming number of AI companies and patents in the US and China, "the AI world is dominated by China and the US" (Zhu 2017).

Offset" strategy, the victory of Google DeepMind's AlphaGo over top-ranked human players in the ancient Chinese board game Go in 2016 shocked China as much as elsewhere. It was a "Sputnik moment" for China that led to some critical Chinese reflection of AI (Mozur 2017). After the Obama administration released three AI reports in late 2016, China put forward its "New Generation AI Development Plan" in July 2017. Not surprisingly, there are striking similarities between American and Chinese AI strategies (Ding 2018) from top-level strategic objectives to specifics regarding policy details and recommendations (Allen and Kania 2017). As such, to Allen and Kania (2017), "China is embracing and implementing America's (AI) strategy". Arguably, the success of American AI has led to considerable anxieties among Chinese strategic analysts and thus pushed China to adopt a national strategic approach to AI.

Domestic Context: AI for Non-Traditional Security

In addition to history and geopolitics, there is also a domestic context for China's securitizing move. AI's impact on China's national security is divided by some Chinese scholars into two aspects: traditional and non-traditional (Que and Zhang 2020; Li 2018). The former refers to the military threat such as the aforementioned use of AI in warfare, while the latter includes non-military sources, such as political security, economic security, environmental security, cybersecurity and energy security (Que and Zhang 2020).Above all, the most important is so-called political security (政治安全) or institutional security (制度安全), i.e. regime security.

As far as regime security is concerned, China's controversial and bold AI practices in state governance is an inevitable topic. As part of the CCP's adaptation strategy in the digital age, China has heavily invested in AI technologies to move towards digital governance.This AI investment expects returns not only in improving public services (by enhancing efficiency) but also in maintaining the authoritarian rule, as the next chapter will discuss. One of the most widely discussed aspects among international analysts is how AI empowers digital surveillance. AI has been used to upgrade China's sophisticated state surveillance programme with the potential to reshape state-society relations, as will be discussed in the next chapter. While similar—though less intensive and extensive—AI surveillance programmes have been implemented worldwide, considerable social resistance has taken place in Western societies to balance states' use of AI due to privacy concerns.[5] In China, however, there is little legal constraint in the relevant AI practices. For example, China has been pioneering AI facial recognition technology, which has been restricted or even banned in many Western societies. In the meanwhile, in order to reduce social resistance, the Chinese government has been actively guiding public opinion towards AI by framing it as positive and modern social progress with enormous benefits of securing public safety, as will be discussed later.

Some Chinese scholars also link AI with economic security. According to Li Zheng from the China Institutes of Contemporary International Relations, for example, the core of China's economic security is to uphold the current socialist market economy with Chinese characteristics (Li 2018). In essence, this economic security is also after regime security—i.e. to secure the CCP's rule. Despite its quasi-capitalist market reforms, the CCP's economic policies have always been constrained by its ideological commitment to being a communist party (Zeng 2015). Thus, the CCP has to uphold some sort of socialist responsibilities for its political legitimacy. Li argues that AI will be able not only to improve socioeconomic governance including market supervision and fighting economic crime but also to strengthen the CCP's capability to manage the macro and microeconomy (Li 2018).

This view echoes the discussion over AI's implications for state-market relations. A key problem of the Soviet-style planned economy is that human central planners are not able to efficiently process and react to market information and thus the system is always inefficient. With the blessing of AI, a super-intelligent computing system could be developed to accurately predict the trend of market forces, making advanced planning possible. If realized, AI may not only significantly upgrade China's Soviet-style central planning system and thus enhance its economic—and ideological—security but also produce a powerful digital technocracy (Araya 2019).

---

[5] In the US, for example, leading tech giants such as Amazon and Microsoft have limited the American police's use of their facial recognition technology (BBC 2020). The social pressure has also led to a growing number of American cities such as San Francisco and Boston to ban the use of facial recognition technology by police and city agencies. In Europe, driven by concern about privacy, the EU put forward a new data privacy law—General Data Protection Regulation—to regulate the transfer of personal data in 2016. Some argue that this strict law would lead to serious negative consequences on AI's use and development in Europe (N. Wallace and Castro 2018).

Needless to say, this discussion inevitably touches upon AI's ideological implications. China's authoritarian values are clearly embedded into its AI practices, as will be discussed in the next chapter, but China is not alone. Many other countries have been building ideological values into their AI development as well.[6] Trump's national AI strategy—"the American AI Initiative"—for example, was keen to develop "AI with American values" with reference to "freedom, guarantees of human rights, the rule of law, stability in our institutions, rights to privacy, respect for intellectual property, and opportunities to all to pursue their dream" (US 2019). Some analysts consider "build democracy into AI" necessary to "wrest control from China" (Webb 2019). Thus, it is argued that ideological competition underpins the global AI race that the US cannot afford to lose (Webb 2019). In this regard, AI is about ideological (in)security in China as much as elsewhere.

Lastly, cybersecurity is worthy of mention here. In contrast to the aforementioned political, economic and ideological securities, technical expertise plays a critical role in cybersecurity discourse. Professional knowledge and skills that are not available to the general public and even security studies scholars grant computer scientists legitimacy and authority within the relevant discourse (Hansen and Nissenbaum 2009). As such, technocratic interpretation is crucial in politics of insecurity (Huysmans 2006: 9). This is relevant to what Hansen and Nissenbaum (2009) introduced as a technification process. It not only heavily relies on technical expertise for its resolution but also simultaneously reads "a politically and normatively neutral agenda that technology serves" and thus de-politicizes the securitized issue (Hansen and Nissenbaum 2009: 1167).

Indeed, with or without political discussions on security, the problem of insecurity exists in certain forms "within professional routines and institutional technology and evolve over time according to professional and bureaucratic or institutional requirements" (Huysmans 2006: 8). As far as AI is concerned, the occurrence of a Chinese AI company's large-scale data breach in 2019 is a reminder of security risks. This incident leaked confidential personal information involving up to 2.56 million users (CCTV.com 2019). When mentioning cybersecurity, the Chinese State Council's AI plan points to two aspects. The first is to

[6] In Europe, for example, some are calling to build liberal and democratic values into AI (Wiewiórowski 2020; Mueller-Kaler 2020; Burrows and Mueller-Kaler 2021).

strengthen AI-empowered solutions to improve cybersecurity. Indeed, AI has demonstrated its use in detecting and mitigating cyber threats and thus its value in civilian cyber defence. According to Capgemini's Reinventing Cybersecurity with AI report, without AI-related technologies, 61 per cent of enterprises cannot detect breach attempts (Columbus 2019). AI-related technologies to protect cybersecurity are expected to further grow within the commercial sectors (Columbus 2019).

The second aspect is to strengthen the protection of AI products and system networks. The widely used AI has brought about not only opportunities but also risks. Ironically, even AI-based security protection can become an insecurity problem. Indeed, the aforementioned example of a data breach that occurred in a Chinese AI company whose principal business is to provide AI-based security protection (CCTV.com 2019).The large-scale data leak indicates the unintended security risk brought about by AI's growing application within commercial sectors.

AUDIENCE: SECURITIZATION TO WHOM?

The previous sections establish that the central government is performing a securitizing move. This section argues that the audience of this move includes Chinese domestic political actors, market actors, academic intellectuals and the general public. China's political actors, especially local governments, are the principal audience. It is important to differentiate the Chinese government into the central state and local states here. The former is the *securitizing actor*, while the latter is a key audience of the central state's *securitizing move*. This differentiation is critical to conducting nuanced analyses of China's AI development.

As discussed in Chapter 2, the analyses of international relations often operate on a misguided assumption that considers the Chinese state a unitary actor as they assume the authoritarian regime as highly unified and capable of mobilizing Chinese society to achieve the central government's objectives (Hill 2016). The same mistake has often been made when it comes to analyses of China's AI strategy. As mentioned, many existing AI analyses misguidedly assume that China follows a coherent "nationally concerted" "top-down" command approach to advance AI in order to achieve geopolitical dominance. The relevant analyses are often followed by a call for a similar national approach to AI in order to address this imminent China threat.

As Chapter 2 discussed, the relevant interpretations have exaggerated China's AI advancement and led to unnecessary anxieties among China's near competitors. Thirty years of China Studies scholarship have shown that the authoritarian system in China is fragmented, decentralized (Lieberthal and Oksenberg 1988; Lieberthal 1992; Lieberthal and Lampton 1992; Schurmann 1966) and more recently internationalized (Jones and Zeng 2019; Hameiri and Zeng 2020; Hameiri and Jones 2016; Hameiri et al. 2019; Jones 2019). Since the early 1980s, China's market reforms programmes have developed a Chinese style of federalism in which local states retain a high level of autonomy within their respective jurisdictions while the central state's power is restricted. This means that (a) their interests do not always overlap and (b) the power relations are not simply a top-down command approach but involve multi-level bargaining and political struggle within the system.

For (a), as far as AI is concerned, while the central government has the big picture in mind, such as AI as a strategic industry, local states are primarily driven by regional interests such as a booming regional AI economy—the larger strategic picture is often irrelevant in local economic plans. For example, while the State Council's AI plan frequently links AI with national security, as previously discussed, the local states' AI policies primarily focus on economic aspects with little reference to national security.[7] After all, it is the central state's primary responsibility to protect national security.

In addition, instead of a "nationally concerted" effort, there is a high level of regional competition over factors of production as local states are not driven by the bigger strategic picture but by regional interests, as discussed in Chapter 2. This requires considerable coordination efforts from the central state. For (b), as the power relations are not a simple command but a bargaining approach, the central state has to find ways to motivate domestic actors. In this context, an AI campaign has been launched for domestic mobilization with securitization as part of the efforts. By labelling AI as a security matter, this securitizing move helps to justify the special treatment of the AI industry and win more support from local actors. So, the point to emphasize is that local states are a key audience that the central state has to convince in order to advance its AI agenda.

---

[7] Exceptions include Guangdong. See (Guangdong 2018).

Similarly, China's market actors are a targeted audience to support the central government's AI agenda. After all, the birth of China's booming internet economy is not a product of state actors but market forces. The central state's ambitious AI goals need to rely on the cooperation of Chinese tech companies. For example, the Chinese Ministry of Science and Technology created a so-called "national AI team of China" with the participation of over 15 Chinese tech giants and start-up companies (Yang 2019).Each company was assigned a distinct and strategic AI field to pioneer—for example, Baidu was tasked with autopilot, Alibaba with smart cities and iFlytek with intelligent voice (Yang 2019). With the market actors on board, the central state hopes to take advantage of their expertise for its own ends.

The securitizing move also targets China's intellectual community. China's new policy concepts often serve as slogans to call for intellectual support (Zeng 2020). This applies to AI slogans as well. The state expects the intellectual community not only to inform of its decision-making but also to contribute their research expertise. For example, the State Council's AI plan actively encourages China's scientists to transfer AI innovations for the use in national defence (China 2017).

The securitization process per se also requires considerable intellectual input from social scientists and policy analysts—it needs them to further develop the security discourse to make it more convincing and rigorous. Soon since the central government started the national AI campaign in 2016, for example, many Chinese scholars echoed the call for a security agenda. This has led to a rapid increase in Chinese academic literature studying security and AI. Figure 3.1 shows the number of Chinese academic journal articles with the words "AI" and "security" in the title. The rising trend from 2016 onwards fits in with the overall growth in AI studies in China. Those studies are expected to help the central government develop a more rigorous security logic.

As China's AI campaign involves state propaganda, the general public is also the audience. Public support is very important to the state's AI agenda, despite China's authoritarian system. At the micro level, for example, it will be helpful to address China's shortage of AI talent— estimated to be over 5 million (Hou et al. 2018)—by encouraging more Chinese students to study AI. At the macro level, the technological breakthrough of AI has to rely on Chinese people's data. As AI requires massive data to train and grow, data is the most important factor in successful AI algorithms (Lee 2018: 14). China is privileged in this regard due to its

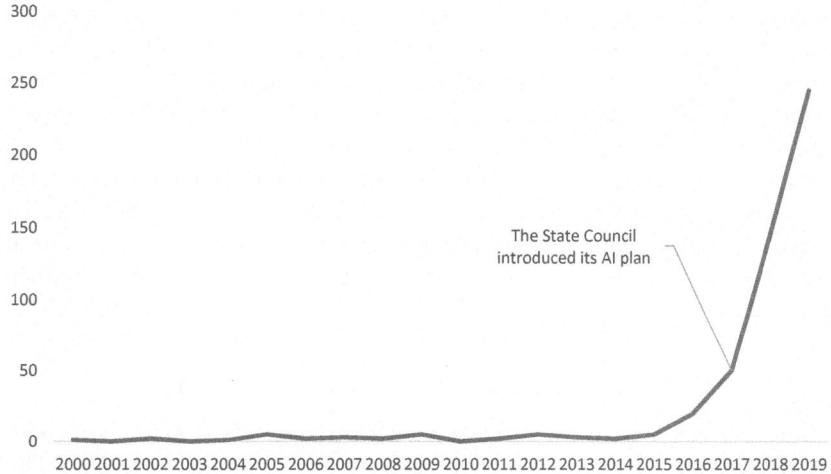

Fig. 3.1 The number of Chinese academic articles with the words "AI" and "security" in the title in China's CNKI database (2001–2019)[8]

huge domestic internet market with over 854 million internet users in 2019 (Xiaoxia 2019),which is twice as many than in the US (Clement 2020).

In order to secure China's data advantage, the government needs a cooperative general public to tolerate the way that their data is being used. In Western democratic societies, the digital use of data has already led to considerable concern about AI's invasion of privacy and thus resistance to its growth, as previously mentioned. The Chinese government hopes the public will buy into its security logic and thus show more tolerance to the negative impact of AI for the sake of national security and better digital service.[9]

Moreover, AI will bring about significant social transformation as will be discussed in the next chapter. During the transition to the age of AI,

[8] The author's brief search from https://www.cnki.net/ on 3 January 2020.

[9] As will be discussed later, state efforts have been made to shape public perceptions regarding AI. The State Council's AI plan, for example, requests domestic governmental organizations to conduct better propaganda work to shape public opinion and social consensus regarding AI in order to prepare for social and ethical challenges brought about by AI technologies (China 2017).

millions of jobs are expected to be replaced by robotics. As McKinsey Global Institute's report shows, by 2030, up to 100 million Chinese workers will need to change their occupations (McKinsey 2017). The massive unemployment and job transition problems that will be brought about by the AI revolution will create potential challenges for China's social stability and thus its authoritarian rule. In this regard, public understanding of the state's AI agenda is critical. In short, despite lack of social resistance and a legal framework to balance the state use of AI, the central government still needs to convince its social actors in order to reduce policy costs.

A Successful or Failed Securitization?

If the central government is making a securitizing move, is it a successful securitization? Currently, securitization is still ongoing in China. The trend of labelling AI as a security matter will or, perhaps more accurately, is becoming more obvious with increasing US-China tensions in technological competition and the wider geopolitical fields. At this stage, it is difficult if not impossible to quantitatively measure the exact impact of the securitizing move because the causal relations are difficult to establish. There is no doubt that local states have got on board with the central government's AI plan and are enthusiastically supporting its AI campaign. Many local AI policies were released immediately after the central state's AI plans in 2016/2017. This is often used by many existing AI analyses as evidence to support their claims about China's nationally concerted approach to AI. However, local governments are primarily driven by the estimated regional gain of a booming AI economy. Many of those provinces had already made their own AI plans before the central state, as discussed in Chapter 2. It is unclear to what extent they are motivated by the securitization move per se rather than the economic benefits brought about by the AI industry.

Economic interests offered by AI play a more obvious role in driving market forces. AI hype is a global phenomenon, and there is no exception in China. The Chinese central state's ambitious AI plan has further contributed to this hype, leading to many concerns about an AI bubble (Chen 2017; Stars 2019). China's market actors have been manipulating the fuzzy definition of AI to redefine AI products, technologies and companies in order to win or even cheat state funding (Deloitte 2018; Wang 2018). In this regard, while market forces are supporting the central government's AI agenda, it is largely a result of potential business interests.

Similarly, it is difficult to quantify how securitization is impacting public opinion. Lack of social resistance to the state use of AI has put China in an advantageous position to grow its AI algorithms. However, this is mostly due to China's authoritarian state's capacity to shape public opinion and national debates. The State Council's AI plan explicitly declares "to guide public opinion" as part of its AI strategy, as will be discussed in the next chapter. Under the call to promote more AI propaganda, positive reporting about AI—such as its contribution to enhance public security including anti-child trafficking (Ye 2019; Zhang 2015) and crime prevention (Yang et al. 2017)—has been widely conducted by the Chinese media. In this regard, the outcome of public cooperation is shaped not only by securitization but also—more importantly—by China's political environment. It remains to be seen how the public will respond when the pain of social transformation brought about by AI, such as massive unemployment, becomes more obvious.

Concluding Remarks: Securitization, so What?

This chapter shows that the Chinese central government is performing a securitizing move by labelling AI as a security matter in order to convince local states, market actors, intellectuals and the general public. If AI is being securitized, so what? Although it is difficult to quantify the exact impact of this securitizing move, it does help the Chinese central government to mobilize domestic actors in order to advance its AI agenda. Despite so, this move also brings about unintended consequences for the securitizing actor's goal in the long run. According to the State Council's AI plan, its three-step AI plan carries specific goals of fostering a booming AI economy and a grand goal of becoming a global AI leader (China 2017). These key objectives may be undermined by securitization for several reasons.

First, a highly securitized AI sector will affect the flow of foreign AI labour and capital to China's AI industry. A booming AI industry requires an outward-looking, open-minded and international socio-politico-economic environment to make it attractive to global talent and capital. However, the securitization trend is pushing in the opposite direction by producing a growing nationalistic inward-looking security discourse of AI. This is counterproductive to China's AI ambitions as it puts China in a disadvantageous position in the global market. More specifically, AI talent is in global shortage, and China is short of over 5 million qualified workers in the AI industry, as previously mentioned. This has led to fierce global competition over qualified labour in the AI industry, and Chinese tech companies have offered very—or unreasonably—high salaries (Chen 2017). This kind of financial attraction, however, could be offset by an unfavourable nationalistic domestic environment.

Second, the securitization trend could hinder economic efficiency. As previously discussed, it contributes to the rise of a self-reliance discourse on technology, which is often made at the expense of economic efficiency. Precisely because China is lagging behind in AI development, it needs to make use of the global AI supply chain to catch up. However, the self-reliance discourse considers the risk of reliance on foreign technology high and thus focuses on "Made in China". This self-reliance is not only difficult (if not impossible) to be realized in the short run but also hinders China's ability to benefit from the global AI market and thus maximize industrial efficiency. Similarly, the securitization of AI in the US also undermines American attraction to Chinese national AI talent, capital and technology. The increasing tension between China and the US has undermined the willingness of Chinese companies to invest in the US. In the long run, it is not desirable to the competitiveness of the American AI industry.

Third, related to the above, this domestic inward-looking nationalistic trend brought about by securitization makes it more difficult for China to realize global leadership. As previously mentioned, China aspires to be a leader in global AI governance and actively advocates to set up international standards and a regulatory framework regarding AI (Allen 2019; Burrows and Mueller-Kaler 2021).[10] Nonetheless, in order to lead AI in the global arena, China needs to provide public goods and win support from others through successful partnerships. It needs to play a key role in promoting global governance, and a global leader needs to act based on common interests not solely on national interests. However, a security-focused inward-looking nationalistic AI discourse is helpful to neither global governance nor common interests. For example, it can contribute to the rise of inward-looking national AI policies that prioritize national

---

[10] In the meanwhile, Chinese scholars have also widely discussed future AI governance and how China can play a leadership role (Gao 2017; Feng 2018).

interests over the globalized world. This contrasts with global governance goals—i.e. to build a shared future through global solidarity. Indeed, many problems brought about by AI such as ethics represent collective challenges to mankind and require a globally concerted response. The inward-looking national AI policies may contribute to a fragmented global governance structure and thus make taking global collective actions to address AI problems more difficult.

As far as global governance is concerned, China's state-centric position is notable. Along with other Brazil, Russia, India, China and South Africa (BRICS) countries, China considers protection of national sovereignty very critical to global governance. As such, it favours a traditional state-centric, sovereignty-oriented regime in global governance.[11] This means that the Chinese view of global AI governance is likely to emphasize the role of national states rather than nongovernmental organizations, Internet giants, and international organizations/regimes. As some Chinese scholars put it:

> sovereign states should be at the top of the AI governance pyramid, with authority to set agendas, formulate policies and reform institutions... in order to face the challenge of information security, it also requires every state to protect national sovereignty and emphasize the role of sovereign states in the development and supervision of AI. (Chen and Yuan 2018)

It remains to be seen how this position will be further developed and contested at the global stage.

Fourth, the securitization trend has been reinforcing technological rivalry at the expense of the potential for global cooperation in AI, and it may further accelerate the US-China confrontation. This is not to deny the existence of US-China cooperation in the field of AI. However, by speaking AI in the language of security and a global race, the relevant security discourse of AI not only emphasizes competition over cooperation and destruction over creation but also requires the creation of a powerful AI rival and perceived threats. This may produce a real security threat—and perhaps a real global AI race—and thus undermines the space for cooperation, which both the US and China can benefit from. In other words, the rivalry discourse adopts a zero-sum angle from geopolitics to

---

[11] For example, see Chinese views of global cyber governance (Zeng et al. 2017) and global economic governance (Zeng 2019).

understanding AI innovation, which inevitably harms the latter in both countries.[12]

More importantly, securitization may further push AI into the area of hard security such as its military application. In order to produce a successful securitization, the securitizing actor tends to exaggerate the security threat. This will enhance the strategic risks of AI's military practices and thus increase the likelihood of war and escalate ongoing conflicts. In this regard, a highly securitized AI politics may set China and the US on a dangerous path towards a catastrophic confrontation that is against everyone's interests and security. In the worst-case scenario, like all other arms races, blithe assertions about the inevitability of AI-enabled war are a self-fulfilling and self-defeating prophecy. In this regard, the aforementioned Fu Ying's call to regulate AI's military application deserves more attention.

Lastly, securitization may undermine the interests of Chinese AI companies by strengthening the state's involvement in China's AI industry. The boundary between the state and the market is already much blurred in China than that in other countries due to China's political environment. By making AI a national security matter, it justifies the necessity of heavier state involvement, if not control. While close ties with the state are a blessing in the Chinese domestic market, they can be a burden on the global stage. Take the aforementioned military-civilian integration as an example. While it helps China's AI companies and research institutes to win more state funding, it undermines their global access given the concerns it has caused among China's AI near-peer competitors. Nowadays, when it comes to AI technology with potential in military use, technological cooperation with China has been increasingly regulated if not banned in Western liberal democratic societies. Some Chinese AI companies including members of the "national AI team of China" have

---

[12] The US-China technology conflict—or the so-called tech cold war—has made both the US and China worse off. Under the Trump administration's pressure, many Chinese telecommunications and semiconductor companies have lost access to not only American but also some European markets. As mentioned, this technology conflict has caused considerable damage to leading Chinese companies such as Huawei and ZTE. This zero-sum geopolitical thinking has already expanded to the AI industry, leading American transaction over a few Chinese AI companies. To American companies, losing sales to China means less funding for research and innovation, which is essential to continuing to secure American leadership in high-tech industries such as semiconductor and AI. It also makes the US less attractive to Chinese capital and AI talent, as previously mentioned.

already been punished by the aforementioned American sanctions due to their close relations with the Chinese government.

In this regard, securitization could hinder China's AI companies' access to the global market and thus future development. It also remains to be seen whether heavier state involvement in the AI industry will hinder market efficiency. Regardless, this kind of state-business relationship is deeply rooted in China's unique socio-political environment and is clearly shaped by China's authoritarian system. The next chapter will look into the development of AI and its implications for China's authoritarian governance.

References

Acharya, Amitav, and Barry Buzan. 2007. "Why Is There no Non-Western International Relations Theory? An Introduction." *International Relations of the Asia-Pacific* 7 (3): 287–312.

Allen, Gregory. 2019. *Understanding China's AI Strategy.* Center for a New American Security. Available at https://www.cnas.org/publications/reports/understanding-chinas-ai-strategy. Accessed on 28 Feburary 2020.

Allen, Gregory, and Elsa Kania. 2017. "China Is Using America's Own Plan to Dominate the Future of Artificial Intelligence." *Foreign Policy.* Available at https://foreignpolicy.com/2017/09/08/china-is-using-americas-own-plan-to-dominate-the-future-of-artificial-intelligence/. Accessed on 28 January 2021.

Alper, Alexandra, Toby Sterling, and Stephen Nellis. 2020. "Trump Administration Pressed Dutch Hard to Cancel China Chip-Equipment Sale: Sources." *Reuters.*

Araya, Daniel. 2019. "Artificial Intelligence and the End of Government." *Forbes.* Available at https://www.forbes.com/sites/danielaraya/2019/01/04/artificial-intelligence-and-the-end-of-government/#678b0efc719b. Accessed on 28 Feburary 2020.

BBC. 2020. "George Floyd: Microsoft Bars Facial Recognition Sales to Police." *BBC News.*

Bo, Yan. 2016. "Securitization and Chinese Climate Change Policy." *Chinese Political Science Review* 1: 94–112.

Burrows, Mathew, and Julian Mueller-Kaler. 2021. *Smart Partnerships amid Great Power Competition: AI, China, and the Global Quest for Digital Sovereignty.* The Atlantic Council GeoTech Center.

Buzan, Barry. 1983. *People, States, and Fear: The National Security Problem in International Relations.* ECPR Press.

Buzan, Barry, Ole Wæver, and Jaap Wilde. 1998. *Security: A New Framework for Analysis*. Lynne Rienner.

Castro, Daniel, Michael McLaughlin, and Eline Chivot. 2019. *Who Is Winning the AI Race: China, the EU or the United States?* Center for Data Innovation. Available at https://www.datainnovation.org/2019/08/who-is-winning-the-ai-race-china-the-eu-or-the-united-states/. Accessed on 28 Feburary 2020.

CCTV.com. 2019. "人工智能企业被曝发生大规模数据泄露事件 超过250万人的数据可被获取 (AI Companies Are Exposed to Large-Scale Data Breaches, Data of More Than 2.5 Million People Are Obtained)." *CCTV.com*.

Chen, Celia. 2017. "China's Artificial Intelligence Sector in Danger of Becoming a 'Bubble', Experts Warn." *South China Morning Post*. Available at https://www.scmp.com/tech/innovation/article/2082217/chinas-artificial-intelligence-sector-danger-becoming-bubble-experts. Accessed on 29 Feburary 2020.

Chen, Weiguang, and Jing Yuan. 2018. "人工智能全球治理: 基于治理主体、结构和机制的分析 (Global AI Governance: An Analysis Based on Governance Subject, Structure and Mechanism)." 国际观察 *(International Review)* 4: 23–37.

China. 2015. "White Paper on China's Military Strategy." Available at http://english.www.gov.cn/archive/white_paper/2015/05/27/content_281475115610833.htm. Accessed on 8 August 2021.

———. 2016a. "互联网+"人工智能三年行动实施方案 *('Internet +' AI Three Years Implementation Plan)*. Available at http://www.gov.cn/xinwen/2016-05/23/content_5075944.htm. Accessed on 19 June 2020.

———. 2016b. 国务院关于印发"十三五"国家战略性新兴产业发展规划的通知 *(Notice of the State Council on Printing and Distributing the Plan for the Development of Strategic Emerging Industries of the State in 13th Five-Year)*. Available at http://www.gov.cn/zhengce/content/2016-12/19/content_5150090.htm. Accessed on 22 January 2021.

———. 2017. 国务院关于印发新一代人工智能发展规划的通知 *(New Generation Artificial Intelligence Development Plan)*. The State Council of China. Available at http://www.gov.cn/zhengce/content/2017-07/20/content_5211996.htm. Accessed on 28 Feburary 2020.

———. 2019. 人工智能具有很强的"头雁"效应 *(Artificial Intelligence Has a Strong "Head Goose" Effect)*. Available at http://paper.people.com.cn/rmrbhwb/html/2019-07/26/content_1938122.htm. Accessed on 28 Feburary 2020.

CISTP. 2018. 中国人工智能发展报告2018 *(2018 Report on China's AI Development)*. Beijing: China Institute for Science and Technology Policy at Tsinghua University (CISTP).

Clement, J. 2020. United States: Number of Internet Users 2000–2019. Available at https://www.statista.com/statistics/276445/number-of-internet-users-in-the-united-states/. Accessed on 31 May 2020.

Columbus, Louis. 2019. "Why AI Is The Future of Cybersecurity." *Forbes*.

Deloitte. 2018. 中国人工智能产业白皮书 *(White Paper on China's AI Industry)*. Deloitte.

Ding, Jeffrey. 2018. *Deciphering China's AI Dream*. Future of Humanity Institute, University of Oxford. Available at https://www.fhi.ox.ac.uk/deciphering-chinas-ai-dream/. Accessed on 28 Feburary 2020.

Ezeokafor, Edwin, and Christian Kaunert. 2018. "Securitization Outside of the West: Conceptualizing the Securitization–Neo-Patrimonialism Nexus in Africa." *Global Discourse: An interdisciplinary Journal of Current Affairs* 8 (1): 83–99.

Feng, Shuai. 2018. "人工智能时代的国际关系:走向变革且不平等的世界 (International Relations in the Age of AI: Moving Towards Shifting and Unequal World)." 外交评论 *(Foreign Affairs Review)* 1: 128–156.

Gao, Qiqi. 2017. "中国在人工智能时代的特殊使命 (China's Special Mission in the Age of AI)." 探索与争鸣 *(Exploration and Contention)* 10: 49–55.

Gledhill, John. 2018. "Securitization, Mafias and Violence in Brazil and Mexico." *Global Discourse: An Interdisciplinary Journal of Current Affairs and Applied Contemporary Thought* 8 (1): 139–154.

Greenwood, Maja, and Ole Wæver. 2013. "Copenhagen–Cairo on a Roundtrip: A Security Theory Meets the Revolution." *Security Dialogue* 44 (5–6): 485–506.

Guangdong. 2018. 广东省人民政府关于印发广东省新一代人工智能发展规划的通知 *(Guangdong's AI Development Plans)*. Guangdong Provincial Government. Available at http://www.gd.gov.cn/gkmlpt/content/0/147/post_147108.html#7. Accessed on 21 June 2020.

Hameiri, Shahar, and Lee Jones. 2016. "Rising Powers and State Transformation: The Case of China." *European Journal of International Relations* 22 (1): 72–98.

Hameiri, Shahar, Lee Jones, and John Heathershaw. 2019. "Reframing the Rising Powers Debate: State Transformation and Foreign Policy." *Third World Quarterly* 40 (8): 1397–1414.

Hameiri, Shahar, and Jinghan Zeng. 2020. "State Transformation and China's Engagement in Global Governance: The Case of Nuclear Technologies." *The Pacific Review* 33 (6): 900–930.

Hansen, Lene, and Helen Nissenbaum. 2009. "Digital Disaster, Cyber Security, and the Copenhagen School." *International Studies Quarterly* 53 (4): 1155–1175.

Hill, Christopher. 2016. *Foreign Policy in the Twenty-First Century*. Basingstoke: Palgrave Macmillan.

Hoadley, Daniel, and Kelley Sayler. 2020. *Artificial Intelligence and National Security*. Congressional Research Service Report. Available at https://fas.org/sgp/crs/natsec/R45178.pdf. Accessed on 3 January 2021.

Holbraad, Martin, and Morten Pedersen. 2012. "Revolutionary Securitization: An Anthropological Extension of Securitization Theory." *International Theory* 4 (2): 165–197.

Hou, Liqiang, Chenglong Jiang, and Fangjie Zhu. 2018. "Competition for Talent Intensifies as China's AI Industry Develops." *China Daily*. Available at http://europe.chinadaily.com.cn/a/201802/05/WS5a77b4aca3106e7dcc13aba2.html. Accessed on 29 Feburary 2020.

Huysmans, Jef. 2006. *The Politics of Insecurity: Fear, Migration and Asylum in the EU*. Routledge.

Ives, Jaqueline, and Anna Holzmann. 2018. *Local Governments Power Up to Advance China's National AI Agenda*. Mercator Institute for China Studies. Available at https://www.merics.org/en/blog/local-governments-power-advance-chinas-national-ai-agenda. Accessed on 28 Feburary 2020.

Jin, Canrong. 2019. "第四次工业革命是中国巨大的历史机遇 (The Fourth Industrial Revolution Is a Huge Historical Opportunity for China)." 北京日报 *(Beijing Daily)*.

Jing, Meng. 2018. *Is Xi Jinping's Iron Grip Better Than Adam Smith's Invisible Hand for Technology Innovation?* Available at https://www.scmp.com/tech/article/2173128/xi-jinpings-iron-grip-better-adam-smiths-invisible-hand-technology-innovation. Accessed on 28 Feburary 2020.

Jones, Lee. 2019. "Theorizing Foreign and Security Policy in an Era of State Transformation: A New Framework and Case Study of China." *Journal of Global Security Studies* 4 (4): 579–597.

Jones, Lee, and Jinghan Zeng. 2019. Understanding China's 'Belt and Road Initiative': Beyond 'grand strategy' to a state transformation analysis. *Third World Quarterly* 40 (8): 1415–1439.

Kania, Elsa. 2020. *"AI Weapons" in China's Military Innovation*. The Brookings Institution. Available at https://www.brookings.edu/research/ai-weapons-in-chinas-military-innovation/. Accessed on 21 June 2020.

Kapur, Saloni. 2018. "From Copenhagen to Uri and Across the Line of Control: India's 'Surgical Strikes' as a Case of Securitisation in Two Acts." *Global Discourse: An Interdisciplinary Journal of Current Affairs and Applied Contemporary Thought* 8 (1): 62–79.

Kapur, Saloni, and Simon Mabon. 2018. "The Copenhagen School Goes Global: Securitisation in the Non-West." *Global Discourse: An Interdisciplinary Journal of Current Affairs and Applied Contemporary Thought* 8 (1).

Kempe, Frederick. 2019. "The US Is Falling Behind China in Crucial Race for AI Dominance." CNBC. Available at https://www.cnbc.com/2019/01/25/

chinas-upper-hand-in-ai-race-could-be-a-devastating-blow-to-the-west.html. Accessed on 28 Feburary 2020.

Krugman, Paul. 2013. "China's Ponzi Bicycle Is Running Into a Brick Wall."

Laliberté, André, and Marc Lanteigne. 2008. "The Issue of Challenges to the Legitimacy of CCP Rule." In *The Chinese Party-State in the 21st Century: Adaptation and the Reinvention of Legitimacy*, edited by André Laliberté and Marc Lanteigne, 1–21. London: Routledge.

Lanier, Jaron, and E. Glen Weyl. 2020. "How Civic Technology Can Help Stop a Pandemic." *Foreign Affairs*. Available at https://www.foreignaffairs.com/articles/asia/2020-03-20/how-civic-technology-can-help-stop-pandemic. Accessed on 20 March 2020.

Lee, Kaifu. 2018. *AI Superpowers: China, Silicon Valley, and the New World Order*. Houghton Mifflin Harcourt.

Lei, Hongzu, Zhimin Zeng, and Shuai Xiong. 2019. "人工智能武器的全球发展、治理风险及对中国的启示 (Implications of AI Weapons for Global Development, Governance Risk and CHINA)." 电子政务 *(E-government)* 203 (11).

Li, Daokui. 2016. "中国会错过第四次工业革命吗 (Will China Miss the Fourth Industrial Revolution?)." 新财富 *(New Fortune)*.

Li, Zheng. 2018. "总体国家安全观视角下的人工智能与国家安全 (Artificial Intelligence and National Security from the Perspective of Overall National Security Concept)." 当代世界 *(Contemporary World)* 10: 18–21.

Lieberthal, Kenneth. 1992. "Introduction: The 'Fragmented Authoritarianism' Model and its Limitations." In *Bureaucracy, Politics and Decision Making in Post-Mao China*, edited by Kenneth Lieberthal and David Lampton, 1–31. Berkeley and London: University of California Press.

Lieberthal, Kenneth, and David Lampton. 1992. *Bureaucracy, Politics, and Decision Making in Post-Mao China*. University of California Press.

Lieberthal, Kenneth, and Michel Oksenberg. 1988. *Policy Making in China*. Princeton University Press.

Liu, Yiling. 2019. "China's AI Dreams Aren't for Everyone." *Foreign Policy*. Available at https://foreignpolicy.com/2019/08/13/china-artificial-intellige nce-dreams-arent-for-everyone-data-privacy-economic-inequality/. Accessed on 28 Feburary 2020.

Mabon, Simon. 2018. "Existential Threats and Regulating Life: Securitization in the Contemporary Middle East." *Global Discourse: An Interdisciplinary Journal of Current Affairs and Applied Contemporary Thought* 8 (1): 42–58.

McKinsey. 2017. *Jobs Lost, Jobs Gained: Workforce Transitions in a Time of Automation*. McKinsey Global Institute.

Merz, Fabien. 2019. "Europe and the Global AI Race." *CSS Analyses in Security Policy* 247.

Mozur, Paul. 2017. "Beijing Wants A.I. to Be Made in China by 2030." *The New York Times.*

Mueller-Kaler, Julian. 2020. *Europe's Third Way.* Atlantic Council. Available at https://www.atlanticcouncil.org/content-series/smart-partnerships/europes-third-way/. Accessed on 16 September 2021.

NISSTC. 2019. 人工智能安全标准化白皮书 *(2019版) (2019 White Paper on AI Standardization).* National Information Security Standardization Technical Committee. Available at https://www.tc260.org.cn/front/postDetail.html?id=20191031151659. Accessed on 21 June 2020.

Nyman, Jonna. 2018. "Securitization." In *Security Studies: An Introduction,* edited by Paul Williams and Matt McDonald. London: Routledge.

Nyman, Jonna, and Jinghan Zeng. 2016. "Securitization in Chinese Climate and Energy Politics." *Wiley Interdisciplinary Reviews: Climate Change* 7 (2): 301–313.

Perry, Elizabeth. 2008. "Chinese Conceptions of "Rights": From Mencius to Mao—And Now." *Perspectives on Politics* 6: 137–147.

Que, Tianshu, and Jiteng Zhang. 2020. "人工智能时代背景下的国家安全治理：应用范式、风险识别与路径选择 (National Security Governance in the Era of Artificial Intelligence: Application Paradigm, Risk Identification and Path Selection)." 国际安全研究 *(Journal of International Security Studies)* 1: 4–38.

Ramanathan, Shriram. 2019. *China's Booming AI Industry: What You Need to Know.* Lux Research. Available at https://www.luxresearchinc.com/blog/chinas-booming-ai-industry-what-you-need-to-know. Accessed on 28 Feburary 2020.

Schurmann, Franz. 1966. *Ideology and Organization in Communist China.* Berkeley and Los Angeles: University of California Press.

Shambaugh, David. 2001. "The Dynamics of Elite Politics During the Jiang Era." *The China Journal* (45).

Stars. 2019. *The Chinese Bubble in Artificial Intelligence Is Insane.* Stars Insights. Available at https://www.the-stars.ch/wp-content/uploads/2019/07/HUANG-Yuanpu_The-Chinese-Bubble-in-Artificial-Intelligence-is-Insane.pdf. Accessed on 29 Feburary 2020.

Trombetta, Maria. 2019. "Securitization of Climate Change in China: Implications for Global Climate Governance." *China Quarterly of International Strategic Studies* 5 (1): 97–116.

US. 2019. *Artificial Intelligence for the American People.* The White House. Available at https://www.whitehouse.gov/ai/. Accessed on 3 January 2021.

Vuori, Juha. 2008. "Illocutionary Logic and Strands of Securitization: Applying the Theory of Securitization to the Study of Non-Democratic Political Orders." *European Journal of International Relations* 14 (1): 65–99.

———. 2011. *How to Do Security with Words: A Grammar of Securitisation in the People's Republic of China.* University of Turku.

Wallace, Nick, and Daniel Castro. 2018. *The Impact of the EU's New Data Protection Regulation on AI*. Center for Data Innovation. Available at https://www2.datainnovation.org/2018-impact-gdpr-ai.pdf. Accessed on 28 January 2021.

Wang, Xinxi. 2018. 伪概念泛滥，认知模糊，AI营销或需一场再教育 (*Pseudo-Concepts Are Proliferated and Cognition Is Fuzzy, AI Marketing May Need a Re-education*). Iyiou. Available at https://www.iyiou.com/p/80682.html. Accessed on 29 Feburary 2020.

Wang, Yuanfeng. 2016. "王元丰: 第四次工业革命真的来了?: (Wang Yuanfeng: Is the Fourth Industrial Revolution Coming?)." *Global Times*.

Wang, Zhengxu. 2005a. "Before the Emergence of Critical Citizens: Economic Development and Political Trust in China." *International Review of Sociology* 15 (1): 155–171.

———. 2005b. "Political trust in China : Forms and Causes." In *Legitimacy: Ambiguities of Political Success or Failure in East and Southeast Asia*, edited by Lynn White. World Scientific Pub Co Inc.

Webb, Amy. 2019. "Build Democracy into AI: Human-Centered Policy is Needed to Wrest Control from China, Tech Giants." *Politico*.

Webster, Graham, Rogier Creemers, Paul Triolo, and Elsa Kania. 2017a. "China's Plan to 'Lead' in AI: Purpose, Prospects, and Problems." *New America*.

———. 2017b. "Full Translation: China's 'New Generation Artificial Intelligence Development Plan' (2017)." *New America*.

Wiewiórowski, Wojciech. 2020. *Artificial Intelligence, Data and Our Values—On the Path to the EU's Digital Future*. European Union. Available at https://edps.europa.eu/press-publications/press-news/blog/artificial-intelligence-data-and-our-values-path-eus-digital_en. Accessed on 3 January 2021.

Wilkinson, Claire. 2007. "The Copenhagen School on Tour in Kyrgyzstan: Is Securitization Theory Useable Outside Europe?" *Security Dialogue* 38 (1): 5–25.

Wood, Peter. 2016. "Chinese Perceptions of the 'Third Offset Strategy'." *China Brief* 16 (15): 1–3.

Xi, Jinping. 2016. "习近平: 为建设世界科技强国而奋斗 (Xi Jinping: Strive to Build a World Technological Power)." *People's Daily*.

Xiaoxia. 2019. "China Has 854 mln Internet Users: Report." *Xinhua*.

Xu, and Naomi. 2019. "The Trump Administration Blacklisted Chinese A.I. Startups. But That Might Not Slow Them Down." *Fortune*.

Yang, Qingqing. 2019. "新一批人工智能"国家队"亮相: 京东、华为、小米等10家企业入选 (A New Batch of AI "National Teams" Debut: JD.com, Huawei, Xiaomi and Other 10 Companies Selected)." *21jingji*.

Yang, Yuan, Yingzhi Yang, and Fei Ju. 2017. "China Seeks Glimpse of Citizens' Future with Crime-Predicting AI."

Ye, Qing. 2019. "人脸识别让寻亲不再是大海捞针 (Face Recognition Makes Searching for Relatives no Longer a Needle in a Haystack)." *Technology Daily*.

Yu, Yifan. 2019. "Why China's AI Players Are Struggling to Evolve Beyond surveillance." *Nikkei Asian Review*. Available at https://asia.nikkei.com/Spotlight/Cover-Story/Why-China-s-AI-players-are-struggling-to-evolve-beyond-surveillance. Accessed on 1 March 2020.

Zeng, Jinghan. 2015. *The Chinese Communist Party's Capacity to Rule: Ideology, Legitimacy and Party Cohesion*. Palgrave Macmillan.

———. 2019a. "Chinese Views of Global Economic Governance." *Third World Quarterly* 40 (3): 578–594.

———. 2020. *Slogan Politics: Understanding Chinese Foreign Policy Concepts*. London: Palgrave Macmillan.

Zeng, Jinghan, and Shaun Breslin. 2016. "China's 'New Type of Great Power Relations': A G2 with Chinese Characteristics?" *International Affairs* 92 (4): 773–794.

Zeng, Jinghan, Tim Stevens, and Yaru Chen. 2017. "China's Solution to Global Cyber Governance: Unpacking the Domestic Discourse of 'Internet Sovereignty'." *Politics and Policy* 45 (3): 432–464.

Zhang, Jingya. 2015. "本市城区郊区城关探头全覆盖 (Probes Fully Cover Our City)." 北京晨报 *(Beijing Morning)*.

Zhang, Tiwei. 2012. "中国不能错过"第三次工业革命" (China Can Not Miss "The Third Industrial Revolution")." 中国青年报 *(China Youth Daily)*.

Zhao, Dingxin. 2009. "The Mandate of Heaven and Performance Legitimation in Historical and Contemporary China." *American Behavioral Scientist* 53 (3): 416–433.

Zhong, Raymond. 2020. "Trump's Latest Move Takes Straight Shot at Huawei's Business." *The New York Times*.

Zhu, Min. 2017. "朱民: 人工智能世界由中美两国主导 需加强人才储备 (Zhu Min: The AI World Is Dominated by China and the United States, and the Talent Pool Needs to Be Strengthened)." *Sina*.

CHAPTER 4

China's Authoritarian Governance and AI

INTRODUCTION

The rise of AI has the potential to transform our governments and societies. An AI revolution may make future governments more digital, efficient and economic than ever before. Yet despite all the benefits that AI offers when applied in governance, Western democratic societies have considerable concerns regarding civil rights. The authoritarian regime in China, on the other hand, has chosen to fully embrace the age of AI. To develop AI is now considered China's national strategy, with a clear goal of making China a leading AI power, as discussed in previous chapters. China's open ambition in AI is considered by many as an open challenge to American AI supremacy (Klein 2020; Allison 2019; Castro et al. 2019; Allen 2019). Following the focus on geopolitical competition brought about by AI in the previous two chapters, this chapter will move on to examine the ideological aspect. Needless to say, China's bold practices of applying AI and big data-related technologies in state governance have further contested the superiority of Western liberal democracy.[1] Some thus call on Europe to stop being "so naive" about China's challenge given the significance of infusing AI algorithms with democratic and liberal values (Burrows and Mueller-Kaler 2021; Mueller-Kaler 2020) and

---

[1] For example, Wright (2019) argues that AI-related technology is empowering China's digital authoritarianism and thus its global competition with liberal democracies.

© The Author(s), under exclusive license to Springer Nature   67
Singapore Pte Ltd. 2022
J. Zeng, *Artificial Intelligence with Chinese Characteristics*,
https://doi.org/10.1007/978-981-19-0722-7_4

the US to "work with its democratic partners" to "build democracy into AI" (Webb 2019).

This chapter examines the Chinese governance approach to AI and big data. It shows that China's bold use of AI practices in governance represents an attempt not only to build a more efficient and capable government to deliver better public services but also to strengthen state control to ensure the continuation of the authoritarian order. The chapter argues that the application of AI and big data-related technology in China's governance should be understood in the wider context of the CCP's broad and incoherent strategy of adaptation to governance by digital means.

The CCP's successful employment of digital technologies to strengthen its governance is made possible by China's unique socio-political environment. China's huge internet market has provided, in effect, unlimited data with which to train and advance AI programmes. As the world shifts from "the age of expertise to the age of data" (Lee 2018: 14), data are a strategic resource that lays the foundation for digital technologies such as AI. As Kaifu Lee, one of the most prominent figures in the Chinese internet sector, points out, data—rather than computing power and AI talent—constitute the most important factor to ensure successful AI algorithms, as "once computing power and engineering talent reach a certain threshold, the quantity of data becomes decisive in determining the overall power and accuracy of an algorithm" (Lee 2018: 14). In 2019, China had over 854 million internet users—and, with this representing only 61.2% of its 1.3 billion population, long-term growth potential (Xiaoxia 2019); by comparison, the US had only 312 million users with less than 10% offline population in which to expand (Clement 2020). In this respect, as the "Saudi Arabia of data", China has considerable comparative advantage in developing its AI industry (Lee 2018: 14).

Moreover, weak civil awareness within Chinese society, combined with the CCP's strong state power, including a well-resourced domestic security sector, has put the Chinese regime in a favourable position not only to exploit its data advantage but also to enhance state control via digital means. For example, China is now leading AI technology in areas such as facial recognition that Europe and the US have put on hold—or even banned—owing to privacy concerns. In contrast to Western democratic societies, for China the key barrier to AI advancement lies in technological rather than legal constraints. More importantly, AI is not just another powerful tech tool to boost China's digital surveillance; AI's automation of decision-making capability without human intervention is unleashing the potential of China's sophisticated digital surveillance network in ways that this chapter will explore. In this respect, the Chinese approach to digital technology has to date been successful in furthering state power and capacity.

The question then arises: will AI and big data-related technology eventually strengthen authoritarian rule or not? While the relevant digital technology has been enhancing the capacity of China's domestic security sector, the AI and big data-powered digital surveillance apparatus alone—no matter how powerful it is—cannot guarantee the continuation of the CCP; and surveillance programmes are only one aspect of the use of AI in China. AI's overall impact is shaped by its interaction with the CCP's key sources of legitimacy, including economic growth, social stability and ideology. Faced with a slowing Chinese economy, the CCP has been counting on a booming high-tech industry to maintain economic growth and thus its legitimacy. Its bold plans for AI were born in this economic context, and the Chinese government has spelled out specific economic targets for its AI industry. In this respect, the economic factor is the most important indicator by which to measure whether China's plans succeed or not. Should China's AI plans deliver a booming AI industry, this will no doubt help the CCP to deliver material goods and thus win popular support.

Yet China's rush towards a booming AI economy entails considerable risks. Indeed, the transition to the age of AI will bring about unprecedented social transformations including but not limited to a restructuring of the workforce. China's proactive push towards an AI economy will only accelerate this process and thus increase the associated risks, while the more cautious approach taken by other countries may mitigate the pain. Given its proactive state approach and its huge population—the world's largest—China will be the country most strongly affected during the transition to the age of AI. Failing to address consequential social problems such as unemployment will threaten China's social order and thus its authoritarian governance. In this respect, while China's AI practices contribute to social stability by empowering the state security apparatus, they may at the same time undermine that stability through the social transformation they will bring about.

Finally, AI seems to be a perfect match for the CCP's ideology. AI has the potential to build super intelligent computing models that can predict market forces without human intervention; this capability may lead to a fundamental reconsideration of our existing perspectives on the flaws of the planned economy and the superiority of the free market. China may evolve into an AI-driven central planning system that maximizes efficiency in allocating market resources. If successful, this will essentially upgrade China's Soviet-style national central planning, producing a powerful digital technocracy with which liberal democracy can hardly compete (Araya 2019).

In addition, with rapid technological development in the long run, AI's efficiency may reach a point where machines can produce abundant material goods and services without the need for any human labour. This will abolish the social contract of "work for a living" that has existed ever since the birth of human society. By then, human society may be entering an ideal utopian world, in which everyone has free and equal access to the distribution of goods, services and capital—the communist society that Karl Marx envisioned in the 1840s. This will lead to a revisiting of the debates on communism vs capitalism and "the end of history". For the CCP's ideological legitimacy, it may be a game-changer. Since the market reforms of the 1980s, the CCP has been suffering from a self-made fundamental contradiction between its declared commitments to socialism and its generation of economic success through quasi-capitalist policies (Zeng 2015). Should AI unlock the potential of central planning and produce sufficient material goods in the remote future, the age of AI will favour the CCP's ideological values and thus its legitimacy. In these respects, the AI revolution is also an ideological revolution.

The Role of Digital Technology in Resilient Authoritarianism

In the late 1980s and early 1990s, the collapse of Eastern European communist regimes and the Soviet Union had worried many including the then CCP leadership. Following the protest of 1989, many including leading China experts predicted the end of the CCP's rule. According to Roderick MacFarquhar, for example, the CCP would soon "follow the CPSU (the Communist Party of the Soviet Union) into the dustbin of history, along with the system it spawned" (MacFarquhar 1991). At the time, Western liberal democracy was hailed as the "final form of human government" and the "end point of mankind's ideological evolution" in Fukuyama's famous "The End of History?" essay (Fukuyama

1989). Scroll forwards three decades, the world seems to be fundamentally different. The internal rise of popularism from the Brexit vote to Trump has badly shaken the moral foundation of liberal democracy. Nowadays, instead of eliminating authoritarianism, liberal democracy has been struggling to cope with the challenges brought about by the rise of authoritarianism in Russia, Saudi Arabia, Iran and—most importantly—China.

The puzzle of why and how the CCP survived instead of following its communist brothers' fate has interested the scholarship (Zeng 2015; Shambaugh 2008). In 2003, Andrew Nathan introduced the concept of "authoritarian resilience" to characterize the Chinese political system (Nathan 2003).[2] Resilient authoritarianism soon became a consensus within the Western China scholarship over the durability of China's authoritarian regime (Pang et al. 2018; Yan 2011). Although increasingly challenged in recent times (Li 2012: 598; Fewsmith and Nathan 2019; Gilley 2003; Baum 2007; Pei 2008; Shirk 2007), "authoritarian resilience" has been widely used to explain authoritarian practices in the Middle East (Haddad 2011, 2012; Heydemann and Leenders 2011), Africa (Haugbølle 2012), Europe (Balmaceda 2014) and Asia (Dukalskis 2016).

The key to resilient authoritarianism lies in the capability and flexibility to respond to internal and external challenges and make the necessary adaptations to accommodate those challenges. This high level of state adaptation capacity ensures the survival of authoritarian order in a dramatically changing socio-political environment such as globalization and digitalization. The rise of information and communication technology (ICT) provides a critical test for this resilience. At the outset, it once raised hopes for the revival of liberal democracy or even the fourth wave of democratization owing to the argument of ICT as a liberating technology.

In the wider context, with the development of the Internet, user-generated content has gradually become dominating. This has led to an academic reflection about the changing nature of information dissemination governance. Some argue that, unlike traditional media, the Internet

---

[2] In his work, Nathan had not clearly defined this concept. Li Cheng (2012: 598) cites Dickson's work to define "authoritarian resilience" as the authoritarian system with the ability to "enhance the capacity of the state to govern effectively through institutional adaptations and policy adjustments".

has empowered society by promoting information flow as it exposes netizens to external foreign ideas that were not available before (Lynch 2011). Thus, it leads to the spread of Western liberal ideas such as democracy and freedom at the cost of pro-authoritarian values. At the domestic level, information flow led by the Internet and ICT also facilities the organization of social protest and opposition forces (Diamond 2010; Lynch 2011; Pierskalla and Hollenbach 2013). In particular, social media sites have attracted considerable academic attention owing to their potential nature as a "liberating technology" that challenges authoritarian rule through collective mobilization.

When analysing the rise of the Arab Spring, for example, some argue that social media played a significant role by facilitating communication among individuals and thus organizing political protests. Not surprisingly, at the time, Egypt's "Facebook Revolution", Syria's "YouTube Uprising" and Iran's "Twitter Uprising" were all hailed as movements in a "social media revolution" (Eltahawy 2010), leading to considerable hope for a fourth wave of democratization (Abushouk 2016; Howard and Hussain 2013). It is argued that the development of the Internet and social media empowers individuals, and fundamentally changes the way in which information is produced, consumed and shared. As a result, this presents a systematic challenge to authoritarian rule, and inevitably undermines the latter (Castells 2009; Benkler 2006; Eltahawy 2010; Shirky 2008). From this emerged the concept of "liberating technology", which argues that modern ICT can deliver liberation to individual citizens by "expand(ing) political, social, and economic freedom" (Diamond 2010: 70).

Critics, however, are sceptical about ICT as a potential liberating factor. Terrorist organizations including Islamic State and other criminal groups, for example, have also taken advantage of social media for sinister goals, instead of liberating purposes. In addition, some argue the view that Arab Spring as a "social media revolution" has overstated the role of ICT (Reardon 2012). The subsequent painful retreat of political transition in the Middle East has led to much reflection on ICT. Needless to say, ICT—and AI—has played an increasingly important role in not only spreading but also combating misinformation, disinformation and fake news. In this regard, ICT is a tool whose effects depend on the context in which it is deployed, and accordingly its function as a driving force of democratization needs to be reconsidered (Shearlaw 2016).

The successful adaptation of ICT to authoritarian contexts has led to an opposite argument, based on the concept not of "liberating technology" but of "repressive technology", according to which these technologies have in fact empowered authoritarian states' capacity to repress civil rights (Rod and Weidmann 2015; Weidmann 2015; Zeng 2016). While facilitating the free flow of information on the one hand, on the other they have given authoritarian regimes both more advanced digital tools with which to *block* this flow and the capacity to shape public opinion by disseminating pro-state views and promoting disinformation campaigns (King et al. 2013, 2014; Zeng 2016; Morozov 2011). In other words, when facing the challenges brought about by ICT, authoritarian regimes can remain resilient by responding appropriately; some have even strengthened their authoritarian governance by mastering ICT to help their cause.

China is the most frequently mentioned and successful case in this regard. The CCP has viewed ICT, AI, big data and other digital technologies, as useful to strengthen its authoritarian rule—its so-called institutional security. China has a proven track record of equipping itself with cutting-edge digital technologies to achieve the so-called "modernization of governance capacity" (Zeng 2016). With the blessing of advanced digital technology, China's heavily invested Golden Shield Project (one of whose subsystems is famously known as the Great Firewall), for example, has strengthened state control on the flow of information and the exposure of Chinese society to foreign ideas (Zeng 2016). This and other such initiatives all reflect the CCP's survival strategy of adapting itself to the digital age and enhancing its governance by electronic means. The CCP's successful digital practices show that resilient authoritarianism can not only cope with the profound challenges brought about by the Internet but also lead the digital trend. This is further demonstrated by the CCP's approach to AI and big data, explored in this chapter.

Big Plans for AI and Big Data: The Top-Level Design

In order to adapt to the digital era, China has made a series of efforts to prepare itself. The Chinese government has adopted a strategic approach and issued a series of policy papers to promote AI and big data-related industry. Big data has been officially announced as an "emerging industry" in China with a series of national policies to support its development. In 2015, for example, the State Council of China issued "the platform for action to promote the development of big data" in order to encourage social innovation and improve governance (China 2015a). According to Chinese Premier Li Keqiang, the Chinese government would make an effort to promote China's "cloud computing" to the international market as it did to China's high-speed rail and nuclear power (Yang 2015).

Following this, a series of high-profile policies on AI were announced. These include the "'Internet+' AI Three Years Implementation Plan", jointly issued by the National Development and Reform Commission, the Ministry of Science and Technology, the Ministry of Industry and Information Technology, and the Cyberspace Administration of China in 2016 (China 2016a), and the "New Generation AI Development Plan", issued by the Chinese State Council in July 2017 (China 2017), as mentioned in previous chapters. The subsequent 19th CCP Congress report echoed the plans in emphasizing the critical role of AI in making China a major manufacturing power in the near future (Xi 2017). As mentioned above, in the relevant official documents, "to vigorously develop AI/big data" has become policy slogans to mobilize domestic actors and thus neither the understanding nor the use of those slogans is coherent. In the meanwhile, the CCP has established the Central Leading Group for Internet Security and Informatization, led by top leaders including Xi Jinping and Li Keqiang in order to embrace the digital era.

China's high-profile digital ambition sits in its national context. As previously mentioned, China has the largest population of mobile phone, internet and social media users in the world. By 2019, China had over 854 million netizens (Xiaoxia 2019). Given that this is only 61.2% of its 1.3 billion population, there is long term growth potential of China's internet and mobile phone user population. As such, there is considerable potential to generate massive data to develop big data-related technology and train cutting-edge AI algorithms in China (Cheng 2014). In this context, the CCP views data as a national strategic resource, and promotes better use of AI and big data as a national strategy, with the hope of unlocking its business potential, as well as improving and securing authoritarian governance in China, as this chapter will explore.

Towards Better Governance?

The Chinese government has always been keen on employing advanced technology to strengthen its governance capacity, and this should be examined in the broader context of the practice of e-government in China. The idea of e-government is to promote a more effective and efficient public service with increased transparency of administrative acts by digitalization. Since the late 1980s, China has been pursuing this e-government strategy by using modern digital technology as part of its modernization programme (Noesselt 2014). With the development of the Internet and social media, the Chinese government has actively adapted its governance strategy to the digital era. For example, Weibo (Chinese version of Twitter) has been adopted into the governance strategy to consult public online opinion in order to rebuild its legitimacy (Noesselt 2014). In this regard, Weibo is used to encourage more political participation and deliberation in the virtual world, and thus strengthen deliberative democracy (Noesselt 2014).

Similarly, the Chinese government is keen to adopt big data-related technology into its governance system. In 2015, the State Council of China issued an official document on how to use big data to improve public governance (China 2015b). This document assigned specific work to governmental departments with a timeline. For example, the Ministry of Commerce, the Administration of Quality Supervision, the Ministry of Industry and Information Technology and other governmental entities were asked to employ big data-related technology to establish a product information traceability system before 2016 (China 2015b).

China's local governments are no less enthusiastic in embracing the opportunities brought about by those digital technologies. Nowadays, they are competing to be pioneers of so-called "intelligent government". Many have already heavily invested in big data. For example, the provincial government of Guizhou has been working with enterprises such as Alibaba to construct a cloud computing infrastructure. Within this cloud services platform, the provincial government shares its data with enterprises and encourages these enterprises to trade their data on the platform. Improving public services is a key goal of this platform. According to an official of Guizhou's Department of Transportation, data integration helps with cooperation between police, fire and health care and thus efficiency has been enhanced 1.5 times by joint duty assignments (Wang 2015). Similarly, by using the cloud services platform to obtain the data about tourism, the government is able to predict the traffic load, the hotel load, perhaps even the security situation and thus be better prepared. Local citizens can also check road traffic and real-time traffic information services by using their phone or iPad to log into Guizhou's Intelligent Transportation Cloud. These local initiatives were clearly supported by central leaders. During Xi Jinping's visit to Guizhou's big data centre, Xi concluded that "I understand it. It is reasonable for Guizhou to develop big data" (Xinhua 2015).

Needless to say, AI-related technology plays a key role in this "intelligent government" vision. Guangzhou municipal government, for example, claimed to be the first in China to introduce facial recognition and "AI + Robot" approval technology into state regulation of the commercial field (Guangzhou 2017). In 2017, Guangzhou Municipal Bureau of Industry and Commerce introduced the "AI + Robot" full electronic commercial registration system. This new system shortens the application process for commercial registration business licences from three days to just ten minutes by employing technologies including facial recognition electronic signature and AI identity verification (Lv and Hu 2017). Following pilot experiments, the model has now been introduced across the entire city of Guangzhou (Guangzhou 2017).

Since 2018, Zhejiang provincial government has been working with Alibaba, using AI to improve its government consultation and complaint reporting platform. It has made use of AI and machine learning to process provincial and city data and establish both a comprehensive provincial government affairs knowledge base and a personalized local government affairs knowledge base (Zhao 2019). In order to provide more diverse and inclusive public services, Zhejiang provincial government has worked with Alibaba to create the first government-affairs-focused AI trainer team in China to facilitate the process (Zhao 2019).

Moreover, the potential of AI in data integration is critical to China's governance, given the huge size of its bureaucracy. As previous chapters discussed, contrary to conventional perception of a highly unified and centralized system, fragmentation has been a key pattern of China's authoritarian regime for decades. Known as a "fragmented authoritarianism model" within the scholarship (Lieberthal 1992; Brødsgaard 2018; Jones and Zeng 2019), this system combines (a) a vertical decentralization, with power and responsibility delegated to different levels of government from central government to provinces, cities, towns and villages, with (b) a horizontal distribution of power among central agencies in Beijing with different but sometimes competing responsibilities.[3] This disjointed pattern of Chinese bureaucracy has allowed high levels of factionalism, localism and departmentalism to emerge. Not surprisingly,

---

[3] The system is called 条条 ("vertical line") and 块块 ("horizontal pieces") in Chinese.

the associated bureaucratic politics has often produced policy outcomes unwelcome to the central government.[4]

Lack of coordination and communication within different governmental organizations is, then, a widespread problem within China's bureaucracy. Massive amounts of data are held by different governmental organizations in islands or silos of information. China's proactive search for AI support carries hopes of improving this fragmented system. An example is the use of AI in integrating state information and upgrading China's surveillance programmes, as will be discussed later.

AI AND BIG DATA-ENABLED SECURITY APPARATUS: THE RISE OF A DIGITAL SURVEILLANCE STATE?

In addition to the above applications, AI and big data-related technology's use in public security is one of if not the most controversial aspects. AI has been a great help to empower the security sector's capacity to fight crime and thus enhance public safety. For example, the Chinese Ministry of Public Security has adopted facial recognition and simulation technology in the fight against child trafficking. The relevant technology solves two key difficulties faced by traditional methods (Ye 2019). The first is that a child's appearance will change significantly over the years after she or he disappears. Using a facial simulation growth algorithm, the relevant technology can help to generate a photo of what a child looks like today based on a photo from his or her childhood (Ye 2019). The second difficulty is identifying lost children from the missing persons' database. After comparing and analysing thousands of photos, it is very easy for human analysts to get confused. AI facial recognition technology, however, significantly enhances not only accuracy but also efficiency: it can reach 99.9% accuracy while making 100,000 facial comparisons per second (Ye 2019). This AI technology has successfully helped thousands of Chinese families to find their lost children (Zhang 2019).

AI is also used to predict crime. Chinese police are working with AI companies to develop a system to assess individuals' chances of committing a crime (Yang et al. 2017). Facial recognition technology and gait analysis are used to monitor individual movements and behaviours, such

---

[4] For examples relating to the BRI, see Jones and Zeng (2019); for examples relating to nuclear governance, see Hameiri and Zeng (2020).

as visits to high-risk places including hardware stores where kitchen knives are sold (Y. Yang, Yang, and Ju 2017). If AI software identifies highly suspicious individuals or groups, it will automatically send warnings to the police (Yang et al. 2017). As Chinese Vice-Minister of Science and Technology Li Meng commented, "if we use our smart systems and smart facilities well, we can know beforehand... who might be a terrorist, who might do something bad" (Yang et al. 2017). This AI application is making *Minority Report*-style policing a reality, and this trend of AI application is not unique to China. American and some European governments, for example, have invested in AI's potential in prediction and counterterrorism, and the relevant technology is used in helping security and intelligence services to assess the risk to airplane passengers and identify terrorist suspects (McKendrick 2019).

In order to ensure the effective operation of those AI and big data digital technologies, access to data is the key. As such, the Chinese government has been strengthening its efforts to obtain private digital information. For example, at the request of the government, Weibo has introduced a real-name registration scheme since 2012 despite the operator's concern about its negative impact. All new Weibo users are required to fill in ID registration, as well as provide their real names, in order to sign up. This registration scheme is also linked with the database of the Ministry of Public Security, which will verify the submitted registration information. The registration is not complete if the name and ID do not match. Thus, inaccurate registration is not allowed. The database of Weibo users has been shared by the police nationwide. The openly declared goal of this scheme is to "regulate the dissemination of objectionable information over the network" (CTCL 2013). The registration system enables the security bureau to track and contain information sources if necessary. Similar moves of registration schemes have been introduced on other digital platforms such as WeChat and Alipay.

By implementing these measures, the regime is able to make individuals in the real world responsible for their behaviours in the virtual world. This has no doubt created a sort of deterrent effect that forces a kind of self-censorship, by which social media users would be extra cautious when posting any sensitive information. In this regard, the administrative regulation on Weibo has affected freedom of speech in virtual space. Notably, the control of freedom occurred almost at the same time when the regime started to use Weibo to consult public online opinion and make itself more responsive to public demands in order to maintain its legitimacy. This indicates a clear strategy towards social media that combines co-optation with coercive control.

In addition to social media, the Chinese government has tightened its control on phone use. Since 2013, the Ministry of Industry and Information Technology has made a new regulation on phone use that requires all telecom services to verify and register user IDs when selling new phone cards (China 2013). AI-related facial recognition technology, for example, has been used to verify the phone card purchaser's identity. With various identity information, the government can track and lock the true identity of phone or Internet users. This enables the regime to contain the information source if there is any, and thus enhances its capacity to crack down on social unrest triggered by petitioners and dissidents. In addition, the regime has attempted to enhance its capacity to forecast large popular protest. As early as 2011, Beijing considered an "Information Platform of Realtime Citizen Movement" system, which would track the precise movement of 17 million mobile phone users in the city (Lewis 2011). This project would provide real-time information about the movement of the population, and thus inform any large-scale social protects.

Moreover, since the 1990s, Chinese governmental organizations have invested heavily in surveillance cameras, developing the largest video surveillance network in the world. This network involves millions of panoramic closed-circuit cameras in public spaces that are working 24 hours a day, 7 days a week. It covers highways, public parks, public transports and taxis, elevators and public streets. Of the world's 10 most watched cities, 8 are in China (Bischoff 2019). In 2018, it was estimated that there was one public camera for every 4.1 Chinese people (Ricker 2019); that may rise to one camera for every 2 people by 2022 (Bischoff 2019). Since 2015, all public streets in Beijing have been monitored by at least 30 million cameras and the 24/7 participation of 4,000 police officers as part of Beijing's Skynet project (Zhang 2015). According to Beijing police, the purpose of these cameras is to prevent "crowd gathering" and street crime (Zhang 2015). The real obstacle to the surveillance scheme is neither a legal obstruction nor social opposition, but environmental pollution—the Haze has significantly undermined the visibility of these cameras, and the regime has to find new technology to allow its cameras to see through the smog (Hall 2013).

Nonetheless, fragmentation within the bureaucracy has contained the full power of China's surveillance programmes. Instead of serving an integrated information network, the monitoring data are fragmented and held in isolation within different departments, which are less committed than many expected to the idea of sharing and coordination. Surveillance cameras, for example, are put in place by a wide range of different governmental departments, public institutions and social organizations (Han 2019). The same street may be watched by dozens of cameras owned by different organizations, leading to a high level of meaningless duplication. In addition, those cameras often have different video standards and information systems, making it difficult to integrate and share the monitoring records (Han 2019). Owing to these duplication and incompatibility problems, China's tremendous investment in video surveillance has not achieved what it could have done. In other words, the power of the world's largest surveillance network has ironically been restrained by fragmented bureaucratic politics.

AI is being employed to overcome these problems. In China's "smart cities" projects, for example, AI technologies have been used to integrate security cameras to break down information isolation within governmental organizations (Han 2019). Cities have piloted experiments to integrate thousands of cameras into city-wide unified video surveillance networks that are capable of having full geographical coverage and 24/7 operation (Han 2019). Here, AI is expected not only to deliver a technological breakthrough but also to raise internal awareness of the need for bureaucratic coordination in order to maximize efficiency.

In addition, some Chinese scholars are discussing the potential of big data and AI-related technology in enabling the regime to track real-time information on the ideological trends of particular groups. The development of the media and the Internet has fundamentally challenged the CCP's ideological indoctrination, as people are exposed to massive amounts of information and the traditional way of information control has become much more ineffective. As such, the CCP views it critical to adapt its ideological indoctrination and political education in the digital era. Some university educators see the rise of big data-related technology as an opportunity to upgrade ideological indoctrination. It is suggested that data mining should focus on the students' digital information (including email, blog, Weibo and WeChat) in order to monitor ideological trends of Chinese college students (Cui 2015).

Some also argue that big data may identify the ideological trend in a timely manner, and thus allow the regime to be more prepared for (if not prevent the occurrence of) the coming crises (Cui 2015). It can also help university educators improve their ability to lead the ideological trend of students. It is argued that ideological indoctrination could occur in a similar way to the improved delivery of online advertisements. For example, Chinese universities could make use of study record data, library book borrowing records and downloads, dissertations and clicks on recent news made by students. By analysing this data, universities may find the focal point of the students, and thus improve their political education accordingly (Hu and Huang 2014).

The Chinese army has also considered big data-related technology as a way of strengthening its political education within the army. For example, an article in *Liberation Army Daily* argues that data is "a valuable resource of education" and suggests establishing a big database to monitor the ideological trend of the army, which will collect data about soldiers' learning and training programmes, online behaviour, communications and liaison as well as their family and social relationships (Lan 2014). It argues that this system will help to increase the effectiveness of political education in the army.

Needless to say, the widely discussed "social credit system"—a national blacklist to assess economic and social reputations of individual citizens and businesses—is also made possible thanks to the development of AI and big data-related technologies. The blueprint for a "social credit system" was introduced by the Chinese government in the 13th Five-Year Plan to strengthen "social management capacity"—a concept that will be discussed later (Yap and Wong 2015). Its design goes beyond Western (mainly American) financial credit rating systems, insofar as it aims to record more than anything financial. Many deemed as "untrustworthy" (失信) in the system are denied access to flights and high-speed railways with some citizens' children are banned from private schools. Many Western analysts and commenters are concerned that the system may score citizens "based on a 'patriotic' criteria such as the content of their postings on social media" (Clover 2016). If so, it may become a tool for the CCP to punish dissenters.

While this is theoretically possible, it is too early to conclude the ultimate form of this system as it is still developing. At the time of writing, this system is still piloting in different Chinese cities, and it has not been integrated into a single unified system across the country yet. For now, this system is suffering from the aforementioned fragmentation and decentralization problems, which AI is supposed to help address. Thus, the impact of a "social credit system" is far from clear. Moreover, while many Western analysts focus on the negative impact of this "state surveillance infrastructure", the public opinion in China is indeed in favour of this system. An empirical study finds that Chinese people are willing to accept the positive benefits—such as encouraging honest behaviour—at the cost of privacy violation brought about by the system (Kostka 2019). In other words, this "social credit system" is viewed as a force for good rather than evil in China.

In addition, despite its recent introduction, this system should not be understood as a brand new project to monitor Chinese citizens and invade their privacy. To some extent, it can be interpreted as an attempt to digitalize China's individual archives (档案) system. Borrowed from the Soviet Union, China's individual archives system records data of individual citizens involving their working experience, political orientation and moral character. When it comes to applications of, for example, postgraduate entrance examinations, state-affiliated job promotion and pension insurance, this individual archives system is often used as key proof material. However, with the growth of the market economy in China, its value has decreased as the private sector does not rely on it to decide, for example, recruitment or promotion. Hence, although it is still crucial to people who work for the government and state-owned enterprises, it has become less relevant for many. The "social credit system" is an attempt to make use of digital technology to modernize and revive the individual archives system (Yap and Wong 2015). In this regard, while it is enhancing state capacity for surveillance, it is more consolidating rather than upgrading how the state monitors and assesses individual citizens.

CHINA'S UNIQUE SOCIO-POLITICAL CONTEXT FOR THE RISE OF AI AND BIG DATA

China's application of AI and big data-related technology in governance has attracted considerable public and media attention across the world. Many tech blogs and analyses have been closely following the development of China's high-tech surveillance state for years. The trend is not unique to China; all capable states are applying the relevant digital technologies in governance, and AI-powered state surveillance has long been making popular headlines. In this regard, the trend of intensification of state surveillance seems inevitable in the digital age. However, the Chinese government's practices are particularly interesting given China's massive investment in AI and big data combined with its authoritarian goals. Its primary need to maintain (if not strengthen) its authoritarian rule has provided a unique Chinese mode of digital governance.

To start with, in the absence of checks and balances by a strong legislative power, China's massive domestic security budget allows its state apparatus to invest in expensive cutting-edge technologies to develop its security forces. After China's emergence from the 2008 financial crisis, the authoritarian regime has become the most adequately resourced national government in the world. China's expenditure on internal security even surpasses its defence spending, despite the fact that China's military budget is the second largest in the world behind the US. Therefore, the regime has the most adequate financial resources to invest in cutting-edge big data technology to equip its security force. This is demonstrated by its heavy investment on the aforementioned "Golden Shield Project".

In seeking to maintain its authoritarian rule, the state has strong incentives to strengthen its control over society, and digital technology is one of its tools. Since 2011, the authoritarian regime has made extra efforts in strengthening its so-called "social management capacity" (Li 2011)— an official concept that refers to social control activities but downplays its coercive connotations—with the hope of constructing a so-called "social management system with Chinese characteristics" (Li 2011). The 18th CCP Congress report used the term "social management" to replace "e-government" (Noesselt 2014: 456). It states that,

> We should improve the online services and advocate healthy themes on the Internet. We should strengthen social management of the Internet and promote orderly network operations in accordance with laws and regulations. We should crack down on pornography and illegal publications and resist vulgar trends. (Noesselt 2014: 456)

Innovation is the key emphasis of "social management capacity" here. The Chinese government has called for all levels of governmental organs to innovate in social management capacity (Li 2011). In this context, with its distinct advantages, digital technology has naturally been adopted into the governance strategy in order to reshape the state-society relations to be in favour of the regime. The state plan includes a grid-style social management model that represents a surveillance system for maintaining public security and social order. The development of AI in social governance is designed to support this broader social management goal

(Hoffman 2019). As the Chinese State Council's "New Generation AI Development Plan" clearly points out:

> AI technologies can accurately sense, forecast, and provide early warning of major situations for infrastructure facilities and social security operations; grasp group cognition and psychological changes in a timely manner; and take the initiative in decision-making and reactions – which will significantly elevate the capability and level of social governance, playing an irreplaceable role in effectively maintaining social stability. (Webster et al. 2017)

More importantly, there is a relatively low level of social resistance to mass surveillance in China (Zeng 2016). It is important to put this into perspective. Even in Western democratic states, where there is supposedly a strong legal framework to balance state power, surveillance programmes such as PRISM can still be implemented by the US National Security Agency to monitor American citizens (Ball 2013). Chinese society, by comparison, has a much weaker awareness of civil rights, and legal resistance to these surveillance programmes is virtually absent. The major obstacle to implementing big data for state surveillance lies in technical aspects rather than legal ones as mentioned. At the same time, China's evolving legal framework pertaining to citizen privacy seems to be erring on the side of the government—in 2015, for example, a national security law was passed to allow the Chinese security bureau full access to the data needed (China 2015c).

The overwhelming authority of the Chinese state grants it the power to control not only legislation but also public opinion. There is an obvious contrast in media narratives about AI between China and Western democratic societies. The public and media discussion over the state use of AI and big data-related technology in the West has often been vigilant to its negative impacts, especially its potential invasion of privacy and harm to civil rights. Those topics, however, have been less discussed in China's public debate—not only because China's censorship grants the state capacity to shape national agendas and debates. Regarding AI, the Chinese state has clearly indicated a strong will to influence public opinion. For example, the Chinese State Council's "New Generation AI Development Plan" includes the objective "to guide public opinion" about AI as part of China's AI strategy (China 2017). According to the plan, China should.

fully use all kinds of traditional media and new media to quickly propagate new progress and new achievements in AI, to let the healthy development of AI become a consensus in all of society, and muster the vigor of all of society to participate in and support the development of AI. Conduct timely public opinion guidance, and respond even better to social, theoretical, and legal challenges that may be brought about by the development of AI. (Webster et al. 2017)

In other words, the state has chosen a "technology for good" narrative regarding AI. Chinese state media has often highlighted positive AI stories such as how advanced AI technology has helped thousands of families to find lost children or assisted the police catch criminals with unprecedented speed. These narratives have helped the state to shape public opinion of AI in its preferred way.

In addition, China's state-society relations have produced a unique commercial ecosystem. Although many of China's internet giants are partly financed by foreign capital and are not owned by the state, this has not prevented the state from receiving their full cooperation—after all, the CCP has party committees in most of those companies as an institutionalized way of influencing their daily operations (Feng 2017). Baidu, the Chinese version of Google, for example, is famous for its close relations with the regime and for following government guidelines such as Internet censorship.

Not surprisingly, Chinese internet giants competed to join the so-called "national AI team of China" certified by the Chinese Ministry of Science and Technology in order to lead the "National AI Open Innovation Platform". Each of the first 5 team members selected was assigned a distinct strategic area to pioneer by the Chinese government—autopilot to Baidu, smart cities to Alibaba, media imaging to Tencent, intelligent voice to iFlytek and intelligent vision to SenseTime (Yang 2019). The team was further expanded in 2018 and 2019 to encompass 10 more Chinese internet giants, including Huawei, Jingdong and Xiaomi, along with a number of start-up companies (Yang 2019). The state's recognition is expected to bring "preferential contract bidding, easier access to finance, and sometimes market share protection" to those private companies (Roberts et al. 2021).

This kind of state-business relations allows China's security forces to have more access to data owned by businesses than their counterparts in Western democracies. At the very least, anything resembling Apple's open refusal to assist the FBI during the FBI-Apple encryption dispute in 2016 (Kharpal 2016) is unlikely to occur in China; neither public opinion nor the law would side with the company, and the price of saying no to the government in China is too high to be contemplated. In 2009, despite its size and global influence, Google's unwilling stance to censor its service at the request of the Chinese government was made at the expense of losing almost the entire Chinese market. Google's painful lessons have demonstrated how helpless an enterprise can be when confronting the state power in China, despite Google's global influence.

Unlike their American counterparts, who have been facing strong social pressure against cooperation with the state security sector,[5] many Chinese AI start-ups are actively working with the state to develop surveillance programmes. In 2019, the Trump administration imposed sanctions on a few Chinese AI start-up companies—some of which were members of the abovementioned "national AI team of China"—in protest over their technical assistance to the Chinese government to strengthen its high-technology surveillance in Xinjiang Province (Swanson and Mozur 2019).

Needless to say, without strong societal checks and balances, the state's digital surveillance programmes have been developing rapidly in China; in the West, meanwhile, society has always played a role in balancing state adoption of those digital technologies, exerting pressure to regulate and thus restrict their development. AI is no exception. Take facial recognition as an example. Its use by police and city agencies has been officially banned in a growing list of American cities including San Francisco, Boston and Oakland.[6] The EU also seriously considered a blanket ban on its use in early 2020 (Espinoza and Murgia 2020). Even Google CEO Sundar Pichai is in favour of a temporary ban (Vincent 2020). Needless to say, a blanket ban on AI facial recognition technology in Europe and the US would give China considerable comparative advantage in the field. Even if the relevant AI technology eventually gains a green light to go ahead in the West, the cautious approach taken by the US and Europe has already given China years in which to advance its AI programmes and thus get ahead. Not surprisingly, China has already taken the lead in facial

[5] For example, under the pressure of the "Black Lives Matter" campaign in 2020, Microsoft joined Amazon and IBM in limiting the use of their facial recognition technology by the US police (BBC 2020).

[6] Other cities where it is banned include San Diego, Berkeley, Somerville, Cambridge and Brookline.

recognition technology. This state of affairs reflects a clear difference in ideological values between China and Western democratic states.

In 2020, however, this ideological gap seems to have become narrower. At the time of writing, the need to combat the coronavirus is changing views on mass surveillance. From South Korea to Italy to Israel, national governments have strengthened their mass surveillance operations to track citizen movements in order to combat the spread of the virus (Singer and Choe 2020). The pandemic has produced a global shift in priorities, favouring the protection of the public interest by preventing the spread of infection at the expense of personal privacy; and if history is anything to go by, this trend may be irreversible. Surveillance programmes designed to combat terrorism, for example, introduced in the wake of 9/11, have remained in place, and the process of strengthening such surveillance continues (Kurra 2011).

Not surprisingly, China is leading the trend. In China, a mobile application called "health code" has been introduced by internet giants Tencent and Alibaba to track users' movements in order to monitor individuals who have already been, or are likely to become, infected with the coronavirus (Mozur et al. 2020). This application shares data with Chinese security and introduces a three-colour dynamic management system, "green, yellow and red", based on individuals' movements and whether they have contacted anyone who has been exposed to the virus (Mozur et al. 2020). A yellow or red code means that the person has a relatively high or high risk of becoming infected, while a green code means low or no risk (Mozur et al. 2020).There have been mandatory checks of this "health code" in the entrances of residential areas, offices, public places and public transport, and only those with green codes can gain access. With nearly a billion Chinese people using this application, problems such as privacy are clearly a concern.

Arguably, the Chinese public is much less sensitive about privacy in comparison with Western society. As co-founder of Baidu Robin Li pointed out in 2018, "the Chinese people are more open or less sensitive about the privacy issue. If they are able to trade privacy for convenience, safety and efficiency, in a lot of cases, they are willing to do that" (Liang 2018). Nonetheless, there is a growing awareness about data protection in China. Indeed, Robin Li's aforementioned view was widely criticized by those who were upset about the way data was mis-used by China's internet giants. The increasing prominence of AI has also contributed to the public discussion over data abuse and privacy issues. Yet most of this discussion focuses on misuse of data by market and private actors rather than state actors. The developing data protection regime is also more focused on how to regulate the former rather than the latter.

In fact, neither AI nor other digital technologies are chief among the causes of China's problems with data leaks and abuse. Even when it comes to offline information, data have often been misused owing to a lack of state regulation and professional conduct. For example, many Chinese phone users regularly receive harassing phone calls selling investment opportunities; there have been many reports of property owners' contact information (stored offline) being sold as a key resource in China's profitable information trafficking industry (Ma et al. 2018). The increasing role of AI in accelerating the collection and centralization of private information has only exacerbated this kind of problem. In 2019, for example, the previously mentioned large-scale data breach occurred in a Chinese AI company focusing on security, leading to the leak of 2.56 million user records including highly sensitive private information such as ID numbers, addresses, dates of birth, photos, work units and location information that can identify users (CCTV.com 2019). The security incident was a big embarrassment to this company, whose principal business is security protection. In this context, the government is facing pressure to regulate information collection in China.

WILL AI STRENGTHEN AUTHORITARIAN RULE BY THE CCP?

Clearly, AI has already been employed to achieve authoritarian ends. But will it eventually strengthen authoritarian rule in China? If focus is entirely on how AI has been applied to upgrade the regime's surveillance capacity and thus its security apparatus, the answer seems to be yes. However, AI is more than surveillance programmes, and its overall impact should be examined more comprehensively. As noted above, this chapter argues that the key point is how China's AI plans interact with the foundations of the CCP's legitimacy. Given that economic growth, social stability and ideology represent three key sources of legitimacy for the CCP (Zeng 2015), it is critically important to understand AI's impact in these three areas in order to understand its overall impact on authoritarian rule in China.

Despite the top-level strategic thinking about a global AI race and AI's use in governance programmes, economic factors remain the fundamental driving force behind China's bold AI plans, as discussed in previous chapters. Economic performance has been the primary source of legitimacy in China since the early 1980s, as will be discussed in detail below. By delivering material outcomes to benefit most Chinese people, the CCP manages to win popular support and stay in power. The recent slowdown of China's economy, however, has led to serious concerns within the national leadership, and many are counting on the expansion of the high-tech sector to save economic growth. This is the thinking behind the bold national strategies launched by China in high-tech development in areas including big data, 5G and AI. Along the same lines, the development of a booming AI economy is the primary goal of China's AI plans. As the Chinese State Council's "New Generation AI Development Plan" points out:

> AI has become a new engine of economic development. AI has become the core driving force for a new round of industrial transformation, [which] will … create a new powerful engine, reconstructing production, distribution, exchange, consumption, etc., links in economic activities … China's economic development enters a new normal, deepening the supply side of structural reform task is very arduous, [and China] must accelerate the rapid application of AI, cultivating and expanding AI industries to inject new kinetic energy into China's economic development. (Webster et al. 2017)

This plan also spells out specific economic goals in a targeted time frame, aiming to reach an AI industry worth more than 150 billion yuan by 2020, 400 billion yuan by 2025 and 1,000 billion yuan by 2030.

It is very important to note that precisely because a booming AI industry is the primary task, China's AI plans are being put into action by the key driving forces of its economy: market, local and subnational actors, not central agencies in Beijing, as discussed in Chapter 2. The hype surrounding AI in China has spurred a great interest in the country's AI plans within the commercial sector, and the provinces have jumped on the bandwagon with the primary aim of boosting their respective regional AI economies. These actors are primarily driven by market and regional interests; the larger geopolitical picture of the global AI race and US-China competition is often irrelevant in their local and practical contexts. Also, the abovementioned fragmentation of Chinese bureaucracy has further complicated the domestic coordination of China's AI practices. This means that the Chinese approach to AI should not be understood as a coherent, nationally concerted effort, and that Beijing's capacity to coordinate China's AI development should not be exaggerated, as previously discussed. Nevertheless, as long as a booming AI industry can be developed to support China's economic growth and thus improve people's living standards, China's success in AI will be a big boost to the CCP's legitimacy.

In comparison, the implications of China's approach to AI for social stability are more complicated. Social stability is another key source of legitimacy for the CCP. In Deng Xiaoping's words, "in China, the overriding need is for stability" (Deng 1994: 284). As discussed above, digital technologies such as AI and big data have empowered the domestic security sector, especially its surveillance capability. Despite being costly, this has strengthened the state's ability to control society. However, repression alone is not sufficient to maintain social stability—not if the state cannot cope with the forthcoming dramatic social changes that AI will bring about.

Among other things, the transition to the age of AI will bring about a fundamental restructuring of the workforce. According to the McKinsey Global Institute, AI will replace up to 30% of the current global workforce by 2030 (McKinsey 2017). Kaifu Lee argues that the main threat of AI is "tremendous social disorder and political collapse stemming from widespread unemployment and gaping inequality", triggering a psychological crisis over the purpose of life and challenging human society's established principle of "working for a living" in a very short period of time (Lee 2018: 21). McKinsey's report forecasts that by 2030, between 400 and 800 million people across the world will need to find jobs in new occupations, and a large proportion of them—up to 100 million or 12% of the 2030 global workforce—will be Chinese, making China the country most strongly affected by the change (McKinsey 2017).

Notably, this social transformation is likely to give China a different experience from that of its near-peer competitors given their market conditions. For example, in the US and Europe, high labour and production costs have provided a strong financial incentive to develop and use AI. In comparison, China has not only cheaper labour costs, as other developing countries do, but also a relatively large population. Many Chinese sectors lack market incentives to apply AI. It remains to be seen how the Chinese government's bold AI push will accelerate this social transformation, and how the associated consequences, including massive unemployment, will pose a critical social challenge for the government.

Although China has some experience of a shifting workforce, having seen the transition from agriculture to industry during its market reforms over the past few decades, some Chinese scholars argue that the unemployment problem looming this time is quite different. According to Gao Qiqi, the director of the AI and Big Data Index Institute at East China University of Political Science and Law, this round of workforce transition will affect not only peasants and workers but also highly educated intellectuals and white-collar workers, who are much more difficult to "soothe and [persuade to] accept the status quo" (Gao 2017). According to Gao,

> The working class has a long history of exercising their pressure in regard to unemployment and thus the society is experienced to face this kind of pressure. However, when facing the unemployment of intellectuals and the tertiary industry, human society has very little experience. This will be a huge challenge for humanity in the future. Therefore, human society is facing a profound revolution with new historical characteristics. (Gao 2017)

In short, AI brings not only advanced digital tools to empower the security sector but also dramatic social changes with which the CCP will have to cope. Whether the CCP can ensure a smooth transition to the age of AI is the key to assessing the impact of AI on China's social stability and thus the regime's legitimacy.

The third and often neglected pillar of the CCP's legitimacy is ideology, and here AI has the potential to be a game-changer. With the marginalization of communism, ideology is widely considered obsolete in China.[7] This is in fact not the case (Zeng 2015). The very purpose of a communist party is supposed to be a vehicle to deliver a communist society. Despite the CCP's remarkable economic success, its market reforms programmes have been questioned by China's leftists, who believe that it is wrong to move Chinese society towards a capitalist path. Thus, the CCP has been facing a fundamental contradiction between its pragmatic use of quasi-capitalist economic policies to generate

---

[7] As Holbig (2013: 61) rightly points out, "in the political science literature on contemporary China, ideology is mostly regarded as a dogmatic straitjacket to market reforms that has been worn out over the years of economic success, an obsolete legacy of the past waiting to be cast off in the course of the country's transition toward capitalism".

economic success and its ideological commitment to communism and socialism (Zeng 2015).

This contradiction can be tracked as far back as the early 1980s, if not the late 1970s. At the time, Mao Zedong's decades-long political campaigns had left behind a failed experiment in communism. The then Soviet-style planned economy—in which the central government had a high level of control over economic activities including production, distribution and allocation—was not just stagnant but broken. While this system ensured a high level of economic equality, it suffered from two key problems: (a) low productivity, as people lacked motivation to work, given that their gain from their labour was not correlated with their input, and (b) low efficiency, as central planners were not able to process and react to information as quickly and efficiently as the market could. In the wider context, many communist regimes, including the Soviet Union, faced the same problem and had to allow the market a bigger role in their economies to address the flaw.

As did China. With Mao's death in 1976, the CCP lost the last source of popular support—the legitimacy deriving from his charisma. In order to save the party, it decided to get rid of the inefficient planned economy and move towards a market economy—named "socialist market economy" to justify the quasi-capitalist reforms under a communist rule (Zeng 2015). Despite the bumpy transition, this move eventually created an economic miracle and saved the party. This market economy has not only significantly enhanced efficiency but also unleashed the Chinese people's incentives for production, providing the fundamental basis for China's economic miracle. This Chinese transition occurred in a wider international context of the universal decline of the Soviet-style planned economy and a wave of communist regime collapses. In the late 1990s, the wider ideological struggle between communism and capitalism that had underlain the Cold War seemed no longer to exist, and victory belonged to capitalism and the free market.[8]

Nonetheless, China's remarkable economic growth has created many side-effects and exposed the flaws of capitalism. Rapid economic growth has led to fundamental changes in China's economic equality. The previous fairly even wealth distribution under Mao has been replaced by a huge gap between the rich and the poor, leading to serious internal

---

[8] Although the rise of BRICS indicates that there is still a debate over to what extent the state should play a role in socioeconomic affairs (Beeson and Zeng 2018).

concerns about the CCP's legitimacy (Zeng 2015). It is ironic to see the success of a communist party in leading quasi-capitalist reforms and creating a high level of economic inequality. Xi Jinping's ideological push on Marxism-Leninism from 2012 onwards should be seen in this context as an initiative undertaken to save the CCP's ideological legitimacy. In short, without these quasi-capitalist market programmes, the regime cannot deliver the material benefits necessary to keep it in power; however, the closer China moves towards capitalism, the bigger the legitimacy crisis the CCP faces. This fundamental contradiction has lain at the core of Chinese politics since the 1980s.

In the age of AI, fulfilling the CCP's socialism commitments *and* delivering economic growth no longer seems to be impossible. AI may provide a solution to the dilemma by fixing the fundamental flaws of the planned economy identified above—low productivity and low efficiency. With the development of AI, future smart machines may achieve a high level of productivity that not only outperforms human labour but also produces tremendous wealth for human society. In this context, people will no longer need to work for a living. If machines can create sufficient material benefits to satisfy the needs of human society, people's lack of work motivation will not be a problem. As Richard Liu, a leading Chinese internet entrepreneur and founder of Jingdong, has pointed out:

> We in China propose communism. In the past, many people thought communism was completely unattainable. Following [developments in] our technological situation over the past two to three years, however, I suddenly realize that communism can be truly achieved in our generation. As robots can do all your work and have already created enormous wealth, the government can distribute it to everyone. There is neither rich nor poor people. All companies can be completely nationalized. China will only need one e-commerce company, one sales company, and [communism] can be achieved. No one will go to work for material ends, and most will struggle for spiritual, emotional [things]. Humans can enjoy or do something artistic and philosophical. (Sina 2017)

The ideal, utopian world heralded by Liu matches the Marxist vision of a communist society that the CCP claims to be its ultimate goal.[9] With

---

[9] Richard Liu's words sparked considerable controversy—not only because his wealth was seen as incompatible with communism, but also because his comments on the nationalization of all companies and realization of communism were made in a sensitive political

the blessing of modern technology, especially AI, human society may have both tremendous wealth and economic equality. This will fundamentally change the way in which governments and societies operate.

Moreover, super AI has the potential to fix the low efficiency problem of the planned economy when it comes to decision-making. While human central planners—no matter how skilful and knowledgeable they are—can never react more quickly and efficiently to information than the market itself, intelligent computing systems can. Super AI, with greater memory and ability to process and analyse information ever faster, is expected to outperform human intelligence in all areas, including decision-making. A super-intelligent computing system may accurately predict trends in market forces and process information with unprecedented speed to plan ahead. This "market-based, plan-driven" model, powered by super AI, may prove superior to the conventional market-driven model.

Needless to say, this future vision looks very remote. After all, AI researchers and scientists have not even invented general AI yet, and super AI remains hypothetical. However, the existence of the discussion is helping the CCP. During the Chinese market reforms mentioned above, the very ideas of "communism", "central planning" and "planned economy" became dated, used only in a derogatory sense. Many liberals consider them to belong to the dustbin of history, while others believe that they are too utopian to be taken seriously. In the context of the discussion about AI, when communism is put forward as a scenario of a future world and linked with modern technology, these ideas begin to sound less negative.[10] The "dated and unrealistic" concepts may sound not so bad or indeed radical at all. In this respect, the discussion offers propaganda value to the CCP's ideological discourses.

In addition to formal ideology such as communism, AI also interacts with informal ideology—popular ideations to legitimize the authoritarian rule (Zeng 2015). Take the CCP's (in)stability discourse as an example. As previously mentioned, social stability is a key source of political legitimacy in China, and the CCP's capability to maintain a stable social order

climate in which the state was strengthening socialist ideology. At the time, many were quite concerned about the diminishing role of the private sector and entrepreneurs in the Chinese economy.

[10] An opinion piece in *Global Times*, for example, argues that the discussion on communism and future technology provides an opportunity for the Chinese public to better understand and support the realization of communism (Shan 2017).

has contributed to the rise of (in)stability discourse—a popular ideation that China needs its current one-party system to provide social stability, otherwise the social order cannot be maintained and public safety will be ruined (Zeng 2015). The development of AI programmes in state governance has strengthened the discourse that the state is capable of adapting itself to maintain social order. The previously mentioned examples of how AI can be used to fight and predict crime have been framed by state propaganda as a positive tool to protect public safety. In this regard, as long as the state controls the discourse, the development of AI in China's security sector will continue to be perceived as a force for good and thus win popular support in China.

The role of nationalism in China's AI plans is also notable. The current technology competition, especially the US counteractions to contain China's AI innovation, has further strengthened the nationalistic discourse that any reliance on foreign technologies is a security risk and thus China needs to master those key technologies by itself in order to maintain its national integrity and dignity. Historical comparison is also critical. As previously discussed, according to China's historical discourses, technology development decides national fate and China's failure in previous industrial revolutions had led it to fall from a leading power to a weak country. Thus, the fourth industrial revolution is a critical opportunity for China to achieve national revival and return to its rightful position. Given AI's critical role in this revolution, China has to master this technology. In this regard, the Chinese government's ambitious plan to lead China to becoming an AI superpower is no doubt helpful to state propaganda. It frames a capable and visionary CCP leadership that leads China's rise to a technology superpower and thus contrasts with a weak Qing Dynasty that was "bullied by others". In short, the state AI plan echoes nationalistic sentiment and thus enhances its ideological legitimacy.

Concluding Remarks

The AI revolution will bring about tremendous change to human society. In state governance, it has already provided a digital tool with which to improve not only public services but also surveillance programmes, and now the coronavirus outbreak seems to be accelerating an irreversible trend of mass surveillance on a global scale. China has opened its arms to embrace this AI revolution. Its authoritarian regime has made bold moves by investing in AI and big data in order to adapt in the digital age. China's unique socio-political environment and ideological belief has allowed it to become a pioneer in applying AI in governance. While it remains to be seen how this AI and big data-powered security apparatus will evolve, China's practices have shown how digital technology can be used to achieve authoritarian ends.

More importantly, AI offers much more than surveillance programmes. China's AI plan is a full package with the aim of boosting the country's economic growth and raising its global status. As this chapter discusses, the overall impact of this Chinese approach will be determined not by how AI interacts with the security apparatus but by how it affects the CCP's key sources of legitimacy—in other words, whether it can (1) foster a booming AI economy in the short run, (2) maintain a stable social order during the AI revolution and (3) prove the superiority of China's authoritarian and even communist ideological values in the remote future. All of these are challenging tasks accompanied by considerable risks, and it remains to be seen whether the CCP will be sufficiently resilient to cope with the challenges. As such, the adaptation of authoritarianism in the age of AI is doomed to be complicated.

REFERENCES

Abushouk, Ahmed. 2016. "The Arab Spring: A Fourth Wave of Democratization?" *DOMES Digest of Middle East Studies* 25 (1): 52–69.

Allen, Gregory. 2019. *Understanding China's AI Strategy*. Center for a New American Security. Available at https://www.cnas.org/publications/reports/understanding-chinas-ai-strategy. Accessed on 28 Feburary 2020.

Allison, Graham. 2019. "Is China Beating America to AI Supremacy?" *The National Interest*. Available at https://nationalinterest.org/feature/china-beating-america-ai-supremacy-106861. Accessed on 28 Feburary 2020.

Araya, Daniel. 2019. "Artificial Intelligence and the End of Government." *Forbes*. Available at https://www.forbes.com/sites/danielaraya/2019/01/04/artificial-intelligence-and-the-end-of-government/#678b0efc719b. Accessed on 28 Feburary 2020.

Ball, James. 2013. "NSA's Prism Surveillance Program: How It Works and What It Can Do." *The Guardian*.

Balmaceda, Margarita. 2014. "Energy Policy in Belarus: Authoritarian Resilience, Social Contracts, and Patronage in a Post-Soviet Environment." *Eurasian Geography and Economics* 55 (5): 514–536.

Baum, Richard. 2007. "The Limits of Authoritarian Resilience." Conference held in the Center for International Studies and Research, Sciences Po, Paris.

BBC. 2020. "George Floyd: Microsoft Bars Facial Recognition Sales to Police." *BBC News.*

Beeson, Mark, and Jinghan Zeng. 2018. "The BRICS and Global Governance: China's Contradictory Role." *The Third World Quarterly* 39 (10): 1962–1978.

Benkler, Yochai. 2006. *The Wealth of Networks: How Social Production Transforms Markets and Freedom.* Yale University Press.

Bischoff, Paul. 2019. "Surveillance Camera Statistics: Which Cities Have the Most CCTV Cameras?" *Comparitech.*

Brødsgaard, Kjeld. 2018. *Chinese Politics as Fragmented Authoritarianism.* Routledge.

Burrows, Mathew, and Julian Mueller-Kaler. 2021. *Smart Partnerships amid Great Power Competition: AI, China, and the Global Quest for Digital Sovereignty.* The Atlantic Council GeoTech Center.

Castells, Manuel. 2009. *Communication Power.* Oxford University Press.

Castro, Daniel, Michael McLaughlin, and Eline Chivot. 2019. *Who Is Winning the AI Race: China, the EU or the United States?* Center for Data Innovation. Available at https://www.datainnovation.org/2019/08/who-is-winning-the-ai-race-china-the-eu-or-the-united-states/. Accessed on 28 Feburary 2020.

CCTV.com. 2019. "人工智能企业被曝发生大规模数据泄露事件 超过 250 万人的数据可被获取 (AI Companies Are Exposed to Large-Scale Data Breaches, Data of More Than 2.5 Million People Are Obtained)." *CCTV.com.*

Cheng, Jackie. 2014. "Big Data for Development in China." UNDP China Working Paper. Available at http://www.cn.undp.org/content/dam/china/docs/Publications/UNDP%20Working%20Paper_Big%20Data%20for%20Development%20in%20China_Nov%202014.pdf. Accessed on 20 November 2015.

China. 2013.《电话用户真实身份信息登记规定》发布 (Announcement on Regulation of Identity of Phone Users).

———. 2015a. "国务院关于印发促进大数据发展行动纲要的通知 (State Council's Decision on Promoting the Development of Big Data)." Available at http://www.gov.cn/zhengce/content/2015-09/05/content_10137.htm. Accessed on 20 November 2015.

———. 2015b. 国务院办公厅关于运用大数据加强对市场主体服务和监管的若干意见 (The State Council's Decision on Use Big Data to Improve Service and Supervision).

———. 2015c. 授权发布: 中华人民共和国国家安全法 (National Security Law of People's Republic of China).

———. 2016. "互联网+"人工智能三年行动实施方案 (*'Internet +' AI Three Years Implementation Plan*). Available at http://www.gov.cn/xinwen/2016-05/23/content_5075944.htm. Accessed on 19 June 2020.

———. 2017. 国务院关于印发新一代人工智能发展规划的通知 (*New Genera-
tion Artificial Intelligence Development Plan*). The State Council of China.
Available at http://www.gov.cn/zhengce/content/2017-07/20/content_5
211996.htm. Accessed on 28 Feburary 2020.

Clement, J. 2020. United States: Number of Internet Users 2000–
2019. Available at https://www.statista.com/statistics/276445/number-of-
internet-users-in-the-united-states/. Accessed on 31 May 2020.

Clover, Charles. 2016. "China: When Big Data Meets Big Brother." *Financial Times*. Available at http://www.ft.com/cms/s/0/b5b13a5e-b847-11e5-
b151-8e15c9a029fb.html#axzz40G2n0kmE. Accessed on 15 February 2016.

CTCL. 2013. China Telecom 2014 Annual Work Conference Highlights.

Cui, Haiying. 2015. "大数据时代高校网络思想政治教育的价值维度与实现方式
(Value and Practice of Online Political Education in the Big Data Era)." 黑
龙江高教研究 (*Heilongjiang Researches on Higher Education*) 3 (Serial No.
251).

Deng, Xiaoping. 1994. *Selected Works of Deng Xiaoping Vol. 3* (邓小平文选第三
卷). Beijing: Foreign Languages Press.

Diamond, Larry. 2010. "Liberation Technology." *Journal of Democracy* 21 (3):
69–83.

Dukalskis, Alexander. 2016. "North Korea's Shadow Economy: A Force for
Authoritarian Resilience or Corrosion?" *Europe-Asia Studies* 68 (3): 487–507.

Eltahawy, Mona. 2010. "Facebook, YouTube, and Twitter Are the New Tools of
Protest in the Arab World." *The Washington Post*.

Espinoza, Javier, and Madhumita Murgia. 2020. "EU Backs Away from Call for
Blanket Ban on Facial Recognition Tech." *Financial Times*.

Feng, Emily. 2017. "Chinese Tech Groups Display Closer Ties with Communist
Party." *Financial Times*.

Fewsmith, Joseph, and Andrew Nathan. 2019. "Authoritarian Resilience Revisited: Joseph Fewsmith with Response from Andrew J. Nathan." *Journal of
Contemporary China* 28 (116): 167–179.

Fukuyama, Francis. 1989. "The End of History?" *The National Interest* Summer
1989: 3–18.

Gao, Qiqi. 2017. "中国在人工智能时代的特殊使命 (China's Special Mission in
the Age of AI)." 探索与争鸣 (*Exploration and Contention*) 10: 49–55.

Gilley, Bruce. 2003. "The Limits of Authoritarian Resilience." *Journal of
Democracy* 14 (1): 18–26.

Guangzhou. 2017. "广州进入"人工智能+机器人"全程电子化商事登记新时代
(Guangzhou Has Entered the New Age of "AI+ Robot" Full Electronic
Business Registration)." *Guangzhou Daily*.

Haddad, Bassam. 2011. *Business Networks in Syria: The Political Economy of
Authoritarian Resilience*. Stanford University Press.

———. 2012. "Syria, the Arab Uprisings, and the Political Economy of Authoritarian Resilience." In *The Arab Spring*, edited by Clement Henry and Jang Ji-Hyang New York: Palgrave Macmillan.

Hall, John. 2013. "China's CCTV Culture Suffers as Record High Pollution and Smog Levels Render Country's 20 Million Surveillance Cameras Effectively Useless." *Independent*. Available at http://www.independent.co.uk/news/world/asia/chinas-cctv-culture-suffers-as-record-high-pollution-and-smog-levels-render-countrys-20-million-8924572.html#gallery. Accessed on 20 November 2015.

Hameiri, Shahar, Lee Jones, and John Heathershaw. 2019. "Reframing the Rising Powers Debate: State Transformation and Foreign Policy." *Third World Quarterly* 40 (8): 1397–1414.

Han, Xiao. 2019. "让信息流动起来：人工智能与政府治理变革 (Making Information Flow: Artificial Intelligence and Governance Reform)." 社会主义研究 *(Socialism Studies)* 4 (246): 79–86.

Haugbølle, Rikke Hostrup. 2012. "'Vive la grande famille des médias tunisiens' Media Reform, Authoritarian Resilience and Societal Responses in Tunisia." *The Journal of North African Studies* 17 (1).

Heydemann, Steven, and Reinoud Leenders. 2011. "Authoritarian Learning and Authoritarian Resilience: Regime Responses to the 'Arab Awakening'." *Globalizations* 8 (5): 647–653.

Hoffman, Samantha. 2019. "Managing the State: Social Credit, Surveillance, and the Chinese Communist Party's Plan for China." In *Artificial Intelligence, China, Russia, and the Global Order*, edited by Nicholas Wright, 48–54. Maxwell Air Force Base, AL: Air University Press.

Holbig, Heike. 2013. "Ideology After the End of Ideology. China and the Quest for Autocratic Legitimation." *Democratization* 20 (1): 61–81.

Howard, Philip, and Muzammil Hussain. 2013. *Democracy's Fourth Wave?: Digital Media and the Arab Spring*. Oxford University Press.

Hu, Zhongyu, and Liya Huang. 2014. "大数据时代大学生思想政治教育面临的问题及应对 (Problem and Solution of College Students Political Education in the Era of Big Data)." 学校党建与思想教育 *(Party Building and Political Education in Universities)* 484.

Jones, Lee, and Jinghan Zeng. 2019. "Understanding China's 'Belt and Road Initiative': Beyond 'Grand Strategy' to a State Transformation Analysis." *Third World Quarterly* 40 (8): 1415–1439.

Kharpal, Arjun. 2016. "Apple vs FBI: All You Need to Know." CNBC.

King, Gary, Jennifer Pan, and Margaret Roberts. 2013. "How Censorship in China Allows Government Criticism but Silences Collective Expression." *American Political Science Review* 107 (2): 326–343.

———. 2014. "Reverse-Engineering Censorship in China: Randomized Experimentation and Participant Observation." *Science* 345 (6199): 1–10.

Klein, Andrés Ortega. 2020. *The U.S.-China Race and the Fate of Transatlantic Relations Part 1: Tech, Values, and Competition*. The Center for Strategic and International Studies (CSIS). Available at https://www.csis.org/analysis/us-china-race-and-fate-transatlantic-relations. Accessed on 28 Feburary 2020.

Kostka, Genia. 2019. "China's Social Credit Systems and Public Opinion: Explaining High Levels of Approval." *New Media & Society* 21 (7): 1565–1593.

Kurra, Babu. 2011. "How 9/11 Completely Changed Surveillance in U.S." *WIRED*.

Lan, Jun. 2014. "政治教育要适应大数据时代要求 (Political Education Needs to Adapt to the Needs of Big Data Era)." 解放军报 *(Liberation Army Daily)*.

Lee, Kaifu. 2018. *AI Superpowers: China, Silicon Valley, and the New World Order*. Houghton Mifflin Harcourt.

Lewis, Leo. 2011. "China Mobile Phone Tracking System Attacked as 'Big Brother' Surveillance." *The Times*.

Li, Cheng. 2012. "The End of the CCP's Resilient Authoritarianism? A Tripartite Assessment of Shifting Power in China." *China Quarterly* 211: 595–623.

Li, Zhangjun. 2011. "扎扎实实提高社会管理科学化水平 建设中国特色社会主义社会管理体系 (Improve Scientific Level of Social Management, Construct Social Management System with Chinese Characteristics)." *People's Daily*.

Liang, Chenyu. 2018. "Are Chinese People 'Less Sensitive' About Privacy?" *Sixth Tone*.

Lieberthal, Kenneth. 1992. "Introduction: The 'Fragmented Authoritarianism' Model and its Limitations." In *Bureaucracy, Politics and Decision Making in Post-Mao China*, edited by Kenneth Lieberthal and David Lampton, 1–31. Berkeley and London: University of California Press.

Lv, Guang, and Linguo Hu. 2017. "3天变10分钟 广州商事登记实现全程"无人化" (From 3 days to 10 Minutes, Guangzhou Commercial Registration Has Achieved "Unmanned" Operation)." *Xinhua*.

Lynch, Marc. 2011. "After Egypt: The Limits and Promise of Online Challenges to the Authoritarian Arab State." *Perspectives on Politics* 9 (2): 301–310.

Ma, Jing, Yidan Luo, Jiaying You, and Wei Wei. 2018. "调查|谁在给你拨打骚扰电话? (Survey: Who Is Giving You Harassing Calls?)." *The Beijing News*.

MacFarquhar, Roderick. 1991. "The Anatomy of Collapse." *New York Reviews of Books*.

McKendrick, Kathleen. 2019. *Artificial Intelligence Prediction and Counterterrorism*. The Royal Institute of International Affairs (The Royal Institute of International Affairs).

McKinsey. 2017. *Jobs Lost, Jobs Gained: Workforce Transitions in a Time of Automation*. McKinsey Global Institute.

Morozov, Evgeny. 2011. *The Net Delusion: The Dark Side of Internet Freedom*. Philadelphia, PA: Public Affairs Philadelphia.

Mozur, Paul, Raymond Zhong, and Aaron Krolik. 2020. "In Coronavirus Fight, China Gives Citizens a Color Code, With Red Flags." *The New York Times*.

Mueller-Kaler, Julian. 2020. *Europe's Third Way*. Atlantic Council. Available at https://www.atlanticcouncil.org/content-series/smart-partnerships/europes-third-way/. Accessed on 16 September 2021.

Nathan, Andrew. 2003. "Authoritarian Resilience." *Journal of Democracy* 14 (1): 6–17.

Noesselt, Nele. 2014. "Microblogs and the Adaptation of the Chinese Party-State's Governance Strategy." *Governance* 27 (3): 449–468.

Pang, Baoqing, Shu Keng, and Lingna Zhong. 2018. "Sprinting with Small Steps: China's Cadre Management and Authoritarian Resilience." *The China Journal* 80: 68–93.

Pei, Minxin. 2008. *China's Trapped Transition: The Limits of Developmental Autocracy*. Cambridge, MA: Harvard University Press.

Pierskalla, Jan, and Florian Hollenbach. 2013. "Technology and Collective Action: The Effect of Cell Phone Coverage on Political Violence in Africa." *American Political Science Review* 107 (2): 207–224.

Reardon, Sara. 2012. "Was the Arab Spring Really a Facebook Revolution?" *New Scientist*. Available at https://www.newscientist.com/article/mg21428596-400-was-the-arab-spring-really-a-facebook-revolution/. Accessed on 11 April 2016.

Ricker, Thomas. 2019. "The US, Like China, Has About One Surveillance Camera for Every Four People, Says Report." *The Verge*.

Roberts, Huw, Josh Cowls, Jessica Morley, Mariarosaria Taddeo, Vincent Wang, and Luciano Floridi. 2021. "The Chinese Approach to Artificial Intelligence: An Analysis of Policy, Ethics, and Regulation." *AI & Society* 36: 59–77.

Rod, Espen, and Nils Weidmann. 2015. "Empowering Activists or Autocrats? The Internet in Authoritarian Regimes." *Journal of Peace Research* 52 (3): 338–351.

Shambaugh, David L. 2008. *China's Communist Party: Atrophy and Adaptation*. Washington, DC: Berkeley.

Shan, Jie. 2017. "Tycoons Spark Discussion on Realization of Communism." *Global Times*.

Shearlaw, Maeve. 2016. "Egypt Five Years on: Was It Ever a 'Social Media Revolution'?" *The Guardian*.

Shirk, Susan. 2007. *China: Fragile Superpower: How China's Internal Politics Could Derail Its Peaceful Rise*. New York: Oxford University Press.

Shirky, Clay. 2008. *Here Comes Everybody: The Power of Organizing Without Organizations*. Allen Lane.

Sina. 2017. "刘强东: 共产主义将在我们这代实现 公司全部国有化 (Liu Qiangdong: Communism Will Be Realized in Our Generation, All Companies Will Be Nationalized)." *Sina*.

Singer, Natasha, and Sang-Hun Choe. 2020. "As Coronavirus Surveillance Escalates, Personal Privacy Plummets." *New York Times*.

Swanson, Ana, and Paul Mozur. 2019. "U.S. Blacklists 28 Chinese Entities Over Abuses in Xinjiang." *New York Times*.

Vincent, James. 2020. "Google Favors Temporary Facial Recognition Ban as Microsoft Pushes Back." *The Verge*.

Wang, Zhiqiu. 2015. ""云上贵州"好处不只一点 打破壁垒共享互通 (The Benefit of Guizhou on Cloud Is More Than a Little)." *Guizhou Daily*.

Webb, Amy. 2019. "Build Democracy into AI: Human-Centered Policy is Needed to Wrest Control from China, Tech Giants." *Politico*.

Webster, Graham, Rogier Creemers, Paul Triolo, and Elsa Kania. 2017. "Full Translation: China's 'New Generation Artificial Intelligence Development Plan' (2017)." *New America*.

Weidmann, Nils. 2015. "Communication, Technology, and Political Conflict: Introduction to the Special Issue." *Journal of Peace Research* 52 (3): 263–268.

Wright, Nicholas. 2019. "Artificial Intelligence's Three Bundles of Challenges for the Global Order." In *Artificial Intelligence, China, Russia, and the Global Order*, edited by Nicholas Wright. Maxwell Air Force Base, AL: Air University Press.

Xi, Jinping. 2017. *The Chinese Communist Party's 19th Party Congress Report*.

Xiaoxia. 2019. "China Has 854 mln Internet Users: Report." *Xinhua*.

Xinhua. 2015. 习近平考察贵州: 贵州发展大数据确实有道理 (Xi Jinping Visited Guizhou: It Is Reasonable for Guizhou to Develop Big Data).

Yan, Xiaojun. 2011. "Regime Inclusion and the Resilience of Authoritarianism: The Local People's Political Consultative Conference in Post-Mao Chinese Politics." *The China Journal* (66): 53–75.

Yang, Fang. 2015. 李克强"大数据词典": 共享、开放、安全 (Li Keqiang's Big Data Dictionary: Share, Open and Security).

Yang, Qingqing. 2019. "新一批人工智能"国家队"亮相: 京东、华为、小米等10家企业入选 (A New Batch of AI "National Teams" Debut: JD.com, Huawei, Xiaomi and Other 10 Companies Selected)." *21jingji*.

Yang, Yuan, Yingzhi Yang, and Fei Ju. 2017. "China Seeks Glimpse of Citizens' Future with Crime-Predicting AI."

Yap, Chuin-Wei Yap, and Gillian Wong. 2015. "China Wants to Tap Big Data to Build a Bigger Brother." *The Wall Street Journal*. Available at http://blogs.wsj.com/chinarealtime/index.php?p=28059&preview=true. Accessed on 20 November 2015.

Ye, Qing. 2019. "人脸识别让寻亲不再是大海捞针 (Face Recognition Makes Searching for Relatives no Longer a Needle in a Haystack)." *Technology Daily*.

Zeng, Jinghan. 2015. *The Chinese Communist Party's Capacity to Rule: Ideology, Legitimacy and Party Cohesion*. Palgrave Macmillan.

———. 2016. "China's Date with Big Data: Will It Strengthen or Undermine the Authoritarian Rule?" *International Affairs* 92 (6): 1443–1462.

Zhang, Jingya. 2015. "本市城区郊区城关探头全覆盖 (Probes Fully Cover Our City)." 北京晨报 *(Beijing Morning)*.

Zhang, Yan. 2019. "公安部：进一步推广人工智能脸部识别技术进行打拐 (Ministry of Public Security: To Further Promote AI Facial Recognition Technology for Abduction)." *China Daily*.

Zhao, Xiaoyan. 2019. "浙江成立全国首支政务人工智能训练师队伍 (Zhejiang Created the First Government Affairs Focused AI Trainer Team)." *China News (Zhejiang)*.

CHAPTER 5

Towards a Research Agenda for AI with Chinese Characteristics

Nowadays, China's approach to achieve its open ambition of becoming a global AI superpower has drawn considerable attention. Collectively, this book concludes that this is a sophisticated and multifaceted Chinese AI approach with complicated and mixed implications for China's key policy objectives and domestic governance. As this book shows, this approach is neither a "top-down" "nationally concerted" strategy nor a direct translation of the will of Beijing's central agencies in order to achieve a single unified geopolitical objective (if there is any) but a broad policy manifesto, accommodating interests of domestic stakeholders. China's local, subnational and non-state actors are major players in China's AI industry, and their struggle and competition for resources have largely shaped the development of China's AI policies.

In order to mobilize domestic actors, the Chinese central government has turned to labelling AI as a security matter, as this book discusses. This has contributed to the increase in security discourse about China's AI innovation, which builds on elements of great power competition, regime security needs and official narratives about technology. While it is too early to conclude the outcome of this securitization move, this book suggests that the securitization trend has brought about complicated consequences for China's AI industry and policy objectives.

Given AI's potential, the Chinese government has also adopted a proactive approach to explore its use in state governance. This represents

J. Zeng, *Artificial Intelligence with Chinese Characteristics*, https://doi.org/10.1007/978-981-19-0722-7_5

part of the CCP's continued efforts to adapt itself in the digital age. AI along with other digital technologies are considered effective means not only to improve public services but also to consolidate the authoritarian rule. Nonetheless, despite the immediate benefits brought about by those advanced technologies, there are considerable challenges ahead of China's economic, social and ideological transitions to the age of AI.

Given the critical role that China plays in global AI competition and its leadership ambition, the future development of China's AI politics deserves close observation. This book suggests three directions to develop a research agenda for AI with Chinese characteristics. The first is to study the domestic politics of China's AI policy process and national strategy. Arguably, this research direction will be critical to advance understanding of China's AI approach. In the context of China's rise as a global AI player, if not superpower, it will shed light on how to learn from both the success and failure of China's lessons and develop evidence-informed policy-making towards China.

Future studies should not adopt a general international relations approach by treating China as a unitary actor as it cannot capture the domestic complexities of China's AI politics. Instead, a domestic political economy angle will be helpful in examining the role of state and non-state actors in China's national AI strategy. More specifically, future analyses can look into the coordination and competition of domestic actors as well as the consequences of their interactions. A series of questions deserve to be investigated further. For example, with the rapid development of China's AI industry, which state actors emerge as the new main drivers of China's AI policies, who falls and why? Whose interests and agendas are dominating the development of China's AI strategy? What are the main arguments in the policy debate regarding AI? How do domestic—especially local and subnational—actors respond to those AI policies? What are the levels of compliance and non-compliance and why? If the non-compliance level is high, close attention shall be paid to whether there is any agency or mechanism that can be designed to coordinate China's AI strategy and its effects.

In addition, given the commercial and international nature of the AI industry, what is the role of non-state and foreign actors in China's AI politics? At the time of writing, the Chinese government has recently taken a series of regulatory actions and interventions to tighten control over China's tech giants. For example, the state has put forward anti-monopoly enforcement over Tencent and Alibaba in order to maintain financial stability and protect consumer rights. This is part of the state effort to strengthen regulations over digital platforms where there is a lack of regulations except censorship. After all, the rapid development of modern technologies often surpasses state regulatory capacity in most countries including China. Will these efforts successfully help the state to rein in China's mighty tech giants? How will this change the existing state-business relations in China's AI industry?

The above topic is closely related to the second research direction, i.e. China's security politics. It needs to look into securitization of China's AI politics and its implications for China's policy objectives. Indeed, the aforementioned state regulatory actions and interventions are clearly built on a security logic. Arguably, security is the principal goal of anti-monopoly enforcement, as protection of consumer rights is largely about personal data security, and the maintenance of financial stability is related to China's financial security. After all, the state's tightening control over tech giants is for the sake of regime security. In addition, national security logic has become increasingly prominent. Didi Chuxing's IPO in the US stock market, for example, immediately invited state scrutiny and punishment due to concerns over the "risks related to national data security" (Li 2021). Obviously, Didi's share listing in the US has made it more difficult for the Chinese government to ensure the user data held by Didi is sequestered from the rest of the world, especially foreign governments. To the Chinese government, information security here is a matter not only of personal data security but also of national security. The Chinese government seems to be determined to enhance supervision—despite its unintended consequences on, for example, foreign investment.

How will this securitization trend develop in the future? Future studies should look into both the shifting language of this security discourse and their actual effects. What is the underpinning shifting security discourse? To what extent will this security discourse convince its audience and what behaviours have occurred as a consequence? How will this security discourse affect foreign perceptions about China's AI industry and what are the consequences? What is the precise impact of a domestic inward-looking nationalistic security discourse on China's leadership ambition in global AI governance?

Needless to say, this securitization trend should be examined in a global context. After all, from regulatory and intervention action against tech giants to concerns over data held by foreign companies, none of these is a unique Chinese phenomenon. Indeed, China's actions against

Didi only mirrored the Trump administration's national security concern over US consumer data held by Chinese companies such as Tencent and ByteDance. In this regard, Chinese and American securitization seem to reinforce each other. Will this further accelerate US-China tech competition? Is the global AI race prophecy, especially US-China competition, self-fulfilling and setting both countries on a dangerous path?

The third research direction is to investigate the evolving Chinese governance of AI. China's bold AI practices in governance firmly sit in its unique political environment and are considered an ideological challenge to Western liberal democracy. In this regard, China has developed a different path of AI development compared with that in the West. Future studies should continue to observe the CCP's adaptability and resilience in the age of AI. What are AI's immediate benefits to governance capacity of the authoritarian regime? Will those AI and big data-enabled state surveillance projects backfire? How will AI and big data technology reshape China's state-society relations? Existing pilot projects such as the "social credit system" and Golden Shield Project deserve close attention.

In the meanwhile, while embracing the age of AI, how will the Chinese government cope with the forthcoming challenges suggested by this book? To what extent, will/has the Chinese government achieve (d) its growth targets for the AI economy? How will the CCP handle the social transformation brought about by the AI revolution? Can it maintain a stable social order and cope with the problems such as rising socioeconomic inequality, unemployment and value shift? How will AI affect the CCP's ideological legitimacy in the long run? Given the critical role of values in AI algorithms, how will value difference affect global AI cooperation and competition? This research direction will help us to understand not only the ultimate impact of AI on China's authoritarian governance but also the wider value competition on the global stage.

References

Li, Jane. 2021. "China's Big Tech Crackdown Has Opened a New Front: National Security." *Quartz*.

REFERENCES

Abushouk, Ahmed. 2016. "The Arab Spring: A Fourth Wave of Democratization?" *DOMES Digest of Middle East Studies* 25 (1): 52–69.

Acharya, Amitav, and Barry Buzan. 2007. "Why Is There no Non-Western International Relations Theory? An Introduction." *International Relations of the Asia-Pacific* 7 (3): 287–312.

Allen, Gregory. 2019. *Understanding China's AI Strategy*. Center for a New American Security. Available at https://www.cnas.org/publications/reports/understanding-chinas-ai-strategy. Accessed on 28 Feburary 2020.

Allen, Gregory, and Elsa Kania. 2017. "China Is Using America's Own Plan to Dominate the Future of Artificial Intelligence." *Foreign Policy*. Available at https://foreignpolicy.com/2017/09/08/china-is-using-americas-own-plan-to-dominate-the-future-of-artificial-intelligence/. Accessed on 28 January 2021.

Allison, Graham. 2019. "Is China Beating America to AI Supremacy?" *The National Interest*. Available at https://nationalinterest.org/feature/china-beating-america-ai-supremacy-106861. Accessed on 28 Feburary 2020.

Alper, Alexandra, Toby Sterling, and Stephen Nellis. 2020. "Trump Administration Pressed Dutch Hard to Cancel China Chip-Equipment Sale: Sources." *Reuters*.

Araya, Daniel. 2019. "Artificial Intelligence and the End of Government." *Forbes*. Available at https://www.forbes.com/sites/danielaraya/2019/01/04/artificial-intelligence-and-the-end-of-government/#678b0efc719b. Accessed on 28 Feburary 2020.

Ball, James. 2013. "NSA's Prism Surveillance Program: How It Works and What It Can Do." *The Guardian*.

Balmaceda, Margarita. 2014. "Energy Policy in Belarus: Authoritarian Resilience, Social Contracts, and Patronage in a Post-Soviet Environment." *Eurasian Geography and Economics* 55 (5): 514–536.

Baum, Richard. 2007. "The Limits of Authoritarian Resilience." Conference held in the Center for International Studies and Research, Sciences Po, Paris.

BBC. 2020. "George Floyd: Microsoft Bars Facial Recognition Sales to Police." *BBC News*.

Beeson, Mark, and Jinghan Zeng. 2018. "The BRICS and Global Governance: China's Contradictory Role." *The Third World Quarterly* 39 (10): 1962–1978.

Benkler, Yochai. 2006. *The Wealth of Networks: How Social Production Transforms Markets and Freedom.* Yale University Press.

Bischoff, Paul. 2019. "Surveillance Camera Statistics: Which Cities Have the Most CCTV Cameras?" *Comparitech.*

Bo, Yan. 2016. "Securitization and Chinese Climate Change Policy." *Chinese Political Science Review* 1: 94–112.

Breslin, Shaun. 1996. *China in the 1980s: Centre-Province Relations in a Reforming Socialist State.* Basingstoke: Palgrave Macmillan.

Brødsgaard, Kjeld. 2018. *Chinese Politics as Fragmented Authoritarianism.* Routledge.

Burrows, Mathew, and Julian Mueller-Kaler. 2021. *Smart Partnerships amid Great Power Competition: AI, China, and the Global Quest for Digital Sovereignty.* The Atlantic Council GeoTech Center.

Buzan, Barry. 1983. *People, States, and Fear: The National Security Problem in International Relations.* ECPR Press.

Buzan, Barry, Ole Wæver, and Jaap Wilde. 1998. *Security: A New Framework for Analysis.* Lynne Rienner.

Callahan, William. 2016. *China's Belt and Road Initiative and the New Eurasian Order.* Norwegian Institute of International Affairs (Oslo).

Castells, Manuel. 2009. *Communication Power.* Oxford University Press.

Castro, Daniel, Michael McLaughlin, and Eline Chivot. 2019a. *Who Is Winning the AI Race: China, the EU or the United States?* Center for Data Innovation. Available at https://www.datainnovation.org/2019/08/who-is-winning-the-ai-race-china-the-eu-or-the-united-states/. Accessed on 28 Feburary 2020.

———. 2019b. *Who Is Winning the AI Race: China, the EU or the United States?* Center for Data Innovation. Available at https://www.datainnovation.org/2019/08/who-is-winning-the-ai-race-china-the-eu-or-the-united-states/. Accessed on 28 Feburary 2020.

CCTV.com. 2019. "人工智能企业被曝发生大规模数据泄露事件 超过250万人的数据可被获取 (AI Companies Are Exposed to Large-Scale Data Breaches, Data of More Than 2.5 Million People Are Obtained)." *CCTV.com.*

CheetahGlobalLab. 2020. 泡沫挤压: AI行业热度骤降, 基础层投资被忽视 (*Bubble Squeeze: AI Industry Plummets and Infrastructure Investment Is Ignored*). 猎豹全球智库 (Cheetah Global Lab). Available at https://36kr.com/p/5282642. Accessed on 29 Feburary 2020.

Chen, Celia. 2017. "China's Artificial Intelligence Sector in Danger of Becoming a 'Bubble', Experts Warn." *South China Morning Post*. Available at https://www.scmp.com/tech/innovation/article/2082217/chinas-artificial-intelligence-sector-danger-becoming-bubble-experts. Accessed on 29 Feburary 2020.

Chen, Weiguang, and Jing Yuan. 2018. "人工智能全球治理: 基于治理主体、结构和机制的分析 (Global AI Governance: An Analysis Based on Governance Subject, Structure and Mechanism)." 国际观察 (*International Review*) 4: 23–37.

Cheng, Jackie. 2014. "Big Data for Development in China." UNDP China Working Paper. Available at http://www.cn.undp.org/content/dam/china/docs/Publications/UNDP%20Working%20Paper_Big%20Data%20for%20Development%20in%20China_Nov%202014.pdf. Accessed on 20 November 2015.

China. 2013.《电话用户真实身份信息登记规定》发布 (Announcement on Regulation of Identity of Phone Users).

———. 2015a. "White Paper on China's Military Strategy." Available at http://english.www.gov.cn/archive/white_paper/2015/05/27/content_281475115610833.htm. Accessed on 8 August 2021.

———. 2015b. "国务院关于印发促进大数据发展行动纲要的通知 (State Council's Decision on Promoting the Development of Big Data)." Available at http://www.gov.cn/zhengce/content/2015-09/05/content_10137.htm. Accessed on 20 November 2015.

———. 2015c. 国务院办公厅关于运用大数据加强对市场主体服务和监管的若干意见 (The State Council's Decision on Use Big Data to Improve Service and Supervision).

———. 2015d. 授权发布: 中华人民共和国国家安全法 (National Security Law of People's Republic of China).

———. 2016a. "互联网+"人工智能三年行动实施方案 (*'Internet +' AI Three Years Implementation Plan*). Available at http://www.gov.cn/xinwen/2016-05/23/content_5075944.htm. Accessed on 19 June 2020.

———. 2016b. 国务院关于印发"十三五"国家战略性新兴产业发展规划的通知 (*Notice of the State Council on Printing and Distributing the Plan for the Development of Strategic Emerging Industries of the State in 13th Five-Year*). Available at http://www.gov.cn/zhengce/content/2016-12/19/content_5150090.htm. Accessed on 22 January 2021.

———. 2017. 国务院关于印发新一代人工智能发展规划的通知 (*New Generation Artificial Intelligence Development Plan*). The State Council of China.

Available at http://www.gov.cn/zhengce/content/2017-07/20/content_5 211996.htm. Accessed on 28 Feburary 2020.

———. 2019. 人工智能具有很强的"头雁"效应 *(Artificial Intelligence Has a Strong "Head Goose" Effect)*. Available at http://paper.people.com.cn/rmr bhwb/html/2019-07/26/content_1938122.htm. Accessed on 28 Feburary 2020.

CISTP. 2018. 中国人工智能发展报告2018 *(2018 Report on China's AI Development)*. Beijing: China Institute for Science and Technology Policy at Tsinghua University (CISTP).

Clement, J. 2020. United States: Number of Internet Users 2000–2019. Available at https://www.statista.com/statistics/276445/number-of-internet-users-in-the-united-states/. Accessed on 31 May 2020.

Clover, Charles. 2016. "China: When Big Data Meets Big Brother." *Financial Times*. Available at http://www.ft.com/cms/s/0/b5b13a5e-b847-11e5-b151-8e15c9a029fb.html#axzz40G2n0kmE. Accessed on 15 February 2016.

Columbus, Louis. 2019. "Why AI Is The Future of Cybersecurity." *Forbes*.

Corrigan, Jack. 2018. "U.S. Needs a National Strategy for Artificial Intelligence, Lawmakers and Experts Say." *Defense One*.

CTCL. 2013. "China Telecom 2014 Annual Work Conference Highlights."

Cui, Haiying. 2015. "大数据时代高校网络思想政治教育的价值维度与实现方式 (Value and Practice of Online Political Education in the Big Data Era)." 黑龙江高教研究 *(Heilongjiang Researches on Higher Education)* 3 (Serial No. 251).

Dai, Sarah. 2018. "Investor warns of Day of Reckoning for 90 per cent of Chinese AI Start-Ups as Funding Dries Up." *South China Morning Post*.

Deloitte. 2018. 中国人工智能产业白皮书 *(White Paper on China's AI Industry)*. Deloitte.

Deng, Xiaoping. 1994. *Selected Works of Deng Xiaoping Vol. 3* (邓小平文选第三卷). Beijing: Foreign Languages Press.

Diamond, Larry. 2010. "Liberation Technology." *Journal of Democracy* 21 (3): 69–83.

Ding, Jeffrey. 2018. *Deciphering China's AI Dream*. Future of Humanity Institute, University of Oxford. Available at https://www.fhi.ox.ac.uk/deciph ering-chinas-ai-dream/. Accessed on 28 Feburary 2020.

———. 2019a. *"China's Current Capabilities, Policies, and Industrial Ecosystem in AI" Testimony before the U.S.-China Economic and Security Review Commission Hearing on Technology, Trade, and Military-Civil Fusion: China's Pursuit of Artificial Intelligence, New Materials, and New Energy*.

———. 2019b. "The Interests Behind China's Artificial Intelligence Dream." In *Artificial Intelligence, China, Russia, and the Global Order: Technological, Political, Global, and Creative Perspectives*, edited by Nicholas Wright, 43–47. Air University Press.

Dukalskis, Alexander. 2016. "North Korea's Shadow Economy: A Force for Authoritarian Resilience or Corrosion?" *Europe-Asia Studies* 68 (3): 487–507.

Eltahawy, Mona. 2010. "Facebook, YouTube, and Twitter Are the New Tools of Protest in the Arab World." *The Washington Post*.

Espinoza, Javier, and Madhumita Murgia. 2020. "EU Backs Away from Call for Blanket Ban on Facial Recognition Tech." *Financial Times*.

Ezeokafor, Edwin, and Christian Kaunert. 2018. "Securitization Outside of the West: Conceptualizing the Securitization–Neo-Patrimonialism Nexus in Africa." *Global Discourse: An interdisciplinary Journal of Current Affairs* 8 (1): 83–99.

Feng, Emily. 2017. "Chinese Tech Groups Display Closer Ties with Communist Party." *Financial Times*.

Feng, Shuai. 2018. "人工智能时代的国际关系:走向变革且不平等的世界 (International Relations in the Age of AI: Moving Towards Shifting and Unequal World)." 外交评论 *(Foreign Affairs Review)* 1: 128–156.

Fewsmith, Joseph, and Andrew Nathan. 2019. "Authoritarian Resilience Revisited: Joseph Fewsmith with Response from Andrew J. Nathan." *Journal of Contemporary China* 28 (116): 167–179.

Fukuyama, Francis. 1989. "The End of History?" *The National Interest* Summer 1989: 3–18.

Gao, Qiqi. 2017. "中国在人工智能时代的特殊使命 (China's Special Mission in the Age of AI)." 探索与争鸣 *(Exploration and Contention)* 10: 49–55.

Gilley, Bruce. 2003. "The Limits of Authoritarian Resilience." *Journal of Democracy* 14 (1): 18–26.

Gledhill, John. 2018. "Securitization, Mafias and Violence in Brazil and Mexico." *Global Discourse: An Interdisciplinary Journal of Current Affairs and Applied Contemporary Thought* 8 (1): 139–154.

Greenwood, Maja, and Ole Wæver. 2013. "Copenhagen–Cairo on a Roundtrip: A Security Theory Meets the Revolution." *Security Dialogue* 44 (5–6): 485–506.

Guangdong. 2018. 广东省人民政府关于印发广东省新一代人工智能发展规划的通知 *(Guangdong's AI Development Plans)*. Guangdong Provincial Government. Available at http://www.gd.gov.cn/gkmlpt/content/0/147/post_1 47108.html#7. Accessed on 21 June 2020.

Guangzhou. 2017. "广州进入"人工智能+机器人"全程电子化商事登记新时代 (Guangzhou Has Entered the New Age of "AI+ Robot" Full Electronic Business Registration)." *Guangzhou Daily*.

Haddad, Bassam. 2011. *Business Networks in Syria: The Political Economy of Authoritarian Resilience*. Stanford University Press.

———. 2012. "Syria, the Arab Uprisings, and the Political Economy of Authoritarian Resilience." In *The Arab Spring*, edited by Clement Henry and Jang Ji-Hyang New York: Palgrave Macmillan.

Hall, John. 2013. "China's CCTV Culture Suffers as Record High Pollution and Smog Levels Render Country's 20 Million Surveillance Cameras Effectively Useless." *Independent*. Available at http://www.independent.co.uk/news/world/asia/chinas-cctv-culture-suffers-as-record-high-pollution-and-smog-levels-render-countrys-20-million-8924572.html#gallery. Accessed on 20 November 2015.

Hameiri, Shahar, and Lee Jones. 2016. "Rising Powers and State Transformation: The Case of China." *European Journal of International Relations* 22 (1): 72–98.

Hameiri, Shahar, Lee Jones, and John Heathershaw. 2019. "Reframing the Rising Powers Debate: State Transformation and Foreign Policy." *Third World Quarterly* 40 (8): 1397–1414.

Hameiri, Shahar, and Jinghan Zeng. 2020. "State Transformation and China's Engagement in Global Governance: The Case of Nuclear Technologies." *The Pacific Review* 33 (6): 900–930.

Han, Xiao. 2019. "让信息流动起来: 人工智能与政府治理变革 (Making Information Flow: Artificial Intelligence and Governance Reform)." 社会主义研究 *(Socialism Studies)* 4 (246): 79–86.

Hannas, Wm. C., and Huey-meei Chang. 2019. *China's Access to Foreign AI Technology: An Assessment*. Center for Security and Emerging Technology.

Hansen, Lene, and Helen Nissenbaum. 2009. "Digital Disaster, Cyber Security, and the Copenhagen School." *International Studies Quarterly* 53 (4): 1155–1175.

Haugbølle, Rikke Hostrup. 2012. "'Vive la grande famille des médias tunisiens' Media Reform, Authoritarian Resilience and Societal Responses in Tunisia." *The Journal of North African Studies* 17 (1).

Heilmann, Sebastian. 2008a. "From Local Experiments to National Policy: The Origins of China's Distinctive Policy Process." *The China Journal* 59: 1–30.

———. 2008b. "Policy Experimentation in China's Economic Rise." *Studies in Comparative International Development* 43: 1–26.

———. 2009. "Maximum Tinkering Under Uncertainty, Unorthodox Lessons from China." *Modern China* 35 (4): 450–462.

———. 2018. *Red Swan: How Unorthodox Policy-Making Facilitated China's Rise*. Hong Kong: The Chinese University of Hong Kong Press.

Heydemann, Steven, and Reinoud Leenders. 2011. "Authoritarian Learning and Authoritarian Resilience: Regime Responses to the 'Arab Awakening'." *Globalizations* 8 (5): 647–653.

Hill, Christopher. 2016. *Foreign Policy in the Twenty-First Century*. Basingstoke: Palgrave Macmillan.

Hoadley, Daniel, and Kelley Sayler. 2020. *Artificial Intelligence and National Security*. Congressional Research Service Report. Available at https://fas.org/sgp/crs/natsec/R45178.pdf. Accessed on 3 January 2021.

Hoffman, Samantha. 2019. "Managing the State: Social Credit, Surveillance, and the Chinese Communist Party's Plan for China." In *Artificial intelligence, China, Russia, and the Global Order*, edited by Nicholas Wright, 48–54. Maxwell Air Force Base, AL: Air University Press.

Holbig, Heike. 2013. "Ideology After the End of Ideology. China and the Quest for Autocratic Legitimation." *Democratization* 20 (1): 61–81.

Holbraad, Martin, and Morten Pedersen. 2012. "Revolutionary Securitization: An Anthropological Extension of Securitization Theory." *International Theory* 4 (2): 165–197.

Hou, Liqiang, Chenglong Jiang, and Fangjie Zhu. 2018. "Competition for Talent Intensifies as China's AI Industry Develops." *China Daily*. Available at http://europe.chinadaily.com.cn/a/201802/05/WS5a77b4aca3106e7dcc13aba2.html. Accessed on 29 Feburary 2020.

Howard, Philip, and Muzammil Hussain. 2013. *Democracy's Fourth Wave?: Digital Media and the Arab Spring*. Oxford University Press.

Hu, Zhongyu, and Liya Huang. 2014. "大数据时代大学生思想政治教育面临的问题及应对 (Problem and Solution of College Students Political Education in the Era of Big Data)." 学校党建与思想教育 (*Party Building and Political Education in Universities*) 484.

Huysmans, Jef. 2006. *The Politics of Insecurity: Fear, Migration and Asylum in the EU*. Routledge.

International Crisis Group. 2012. "Stirring Up the South China Sea (I)." Asia Report No. 223.

Ives, Jaqueline, and Anna Holzmann. 2018. *Local Governments Power Up to Advance China's National AI Agenda*. Mercator Institute for China Studies. Available at https://www.merics.org/en/blog/local-governments-power-advance-chinas-national-ai-agenda. Accessed on 28 Feburary 2020.

Iyiou. 2018. 19省、市人工智能相关政策研究(上) (*Studies on the Relevant AI Policies of 19 Provinces*). Iyiou. Available at https://www.iyiou.com/intelligence/insight65899.html. Accessed on 29 Feburary 2020.

Jin, Canrong. 2019. "第四次工业革命是中国巨大的历史机遇 (The Fourth Industrial Revolution Is a Huge Historical Opportunity for China)." 北京日报 (*Beijing Daily*).

Jing, Meng. 2018. *Is Xi Jinping's Iron Grip Better Than Adam Smith's Invisible Hand for Technology Innovation?* Available at https://www.scmp.com/tech/article/2173128/xi-jinpings-iron-grip-better-adam-smiths-invisible-hand-technology-innovation. Accessed on 28 Feburary 2020.

Jones, Lee. 2019. "Theorizing Foreign and Security Policy in an Era of State Transformation: A New Framework and Case Study of China." *Journal of Global Security Studies* 4 (4): 579–597.

Jones, Lee, and Jinghan Zeng. 2019. "Understanding China's 'Belt and Road Initiative': Beyond 'Grand Strategy' to a State Transformation Analysis." *Third World Quarterly* 40 (8): 1415–1439.

Jones, Lee, and Yizheng Zou. 2017. "Rethinking the Role of State-owned Enterprises in China's Rise." *New Political Economy* 22 (6): 743–760.

Kania, Elsa. 2017. "China's Artificial Intelligence Revolution: A New AI Development Plan Calls for China to Become the World Leader in the Field by 2030." *The Diplomat.*

———. 2020. *"AI Weapons" in China's Military Innovation.* The Brookings Institution. Available at https://www.brookings.edu/research/ai-weapons-in-chinas-military-innovation/. Accessed on 21 June 2020.

Kaplan, Andreas, and Michael Haenlein. 2019. "Siri, Siri, in My Hand: Who's the Fairest in the Land? On the Interpretations, Illustrations, and Implications of Artificial Intelligence." *Business Horizons* 62: 15–25.

Kapur, Saloni. 2018. "From Copenhagen to Uri and Across the Line of Control: India's 'Surgical Strikes' as a Case of Securitisation in Two Acts." *Global Discourse: An Interdisciplinary Journal of Current Affairs and Applied Contemporary Thought* 8 (1): 62–79.

Kapur, Saloni, and Simon Mabon. 2018. "The Copenhagen School Goes Global: Securitisation in the Non-West." *Global Discourse: An Interdisciplinary Journal of Current Affairs and Applied Contemporary Thought* 8 (1).

Kempe, Frederick. 2019. "The US Is Falling Behind China in Crucial Race for AI Dominance." CNBC. Available at https://www.cnbc.com/2019/01/25/chinas-upper-hand-in-ai-race-could-be-a-devastating-blow-to-the-west.html. Accessed on 28 Feburary 2020.

Kewalramani, Manoj. 2018. *China's Quest for AI Leadership: Prospects and Challenges.* Takshashila Institution.

Kharpal, Arjun. 2016. "Apple vs FBI: All You Need to Know." CNBC.

King, Gary, Jennifer Pan, and Margaret Roberts. 2013. "How Censorship in China Allows Government Criticism but Silences Collective Expression." *American Political Science Review* 107 (2): 326–343.

———. 2014. "Reverse-Engineering Censorship in China: Randomized Experimentation and Participant Observation." *Science* 345 (6199): 1–10.

Klein, Andrés Ortega. 2020. *The U.S.-China Race and the Fate of Transatlantic Relations Part 1: Tech, Values, and Competition.* The Center for Strategic and International Studies (CSIS). Available at https://www.csis.org/analysis/us-china-race-and-fate-transatlantic-relations. Accessed on 28 Feburary 2020.

Koch-Weser, Iacob. 2013. *The Reliability of China's Economic Data: An Analysis of National Output.* The U.S.-China Economic and Security Review Commission. Available at https://www.uscc.gov/sites/default/files/Research/The ReliabilityofChina'sEconomicData.pdf. Accessed on 29 Feburary 2020.

Kostka, Genia. 2019. "China's Social Credit Systems and Public Opinion: Explaining High Levels of Approval." *New Media & Society* 21 (7): 1565–1593.

Krugman, Paul. 2013. "China's Ponzi Bicycle Is Running Into a Brick Wall."

Kurra, Babu. 2011. "How 9/11 Completely Changed Surveillance in U.S." *WIRED*.

Laliberté, André, and Marc Lanteigne. 2008. "The Issue of Challenges to the Legitimacy of CCP Rule." In *The Chinese Party-State in the 21st Century: Adaptation and the Reinvention of Legitimacy*, edited by André Laliberté and Marc Lanteigne, 1–21. London: Routledge.

Lan, Jun. 2014. "政治教育要适应大数据时代要求 (Political Education Needs to Adapt to the Needs of Big Data Era)." 解放军报 *(Liberation Army Daily)*.

Lanier, Jaron, and E. Glen Weyl. 2020. "How Civic Technology Can Help Stop a Pandemic." *Foreign Affairs*. Available at https://www.foreignaffairs.com/articles/asia/2020-03-20/how-civic-technology-can-help-stop-pandemic. Accessed on 20 March 2020.

Laskai, Lorand. 2017. "Beijing's AI Strategy: Old-School Central Planning with a Futuristic Twist." *Council on Foreign Relations*.

Lee, Kaifu. 2018. *AI Superpowers: China, Silicon Valley, and the New World Order*. Houghton Mifflin Harcourt.

Lei, Hongzu, Zhimin Zeng, and Shuai Xiong. 2019. "人工智能武器的全球发展、治理风险及对中国的启示 (Implications of AI Weapons for Global Development, Governance Risk and CHINA)." 电子政务 *(E-government)* 203 (11).

Leverett, Flynt, and Bingbing Wu. 2017. "The New Silk Road and China's Evolving Grand Strategy." *The China Journal* 77 (1): 110–132.

Lewis, Leo. 2011. "China Mobile Phone Tracking System Attacked as 'Big Brother' Surveillance." *The Times*.

Li, Cheng. 2012. "The End of the CCP's Resilient Authoritarianism? A Tripartite Assessment of Shifting Power in China." *China Quarterly* 211: 595–623.

Li, Daokui. 2016. "中国会错过第四次工业革命吗 (Will China Miss the Fourth Industrial Revolution?)." 新财富 *(New Fortune)*.

Li, Jane. 2021. "China's Big Tech Crackdown Has Opened a New Front: National Security." *Quartz*.

Li, Zhangjun. 2011. "扎扎实实提高社会管理科学化水平 建设中国特色社会主义社会管理体系 (Improve Scientific Level of Social Management, Construct Social Management System with Chinese Characteristics)." *People's Daily*.

Li, Zheng. 2018. "总体国家安全观视角下的人工智能与国家安全 (Artificial Intelligence and National Security from the Perspective of Overall National Security Concept)." 当代世界 *(Contemporary World)* 10: 18–21.

Liang, Chenyu. 2018. "Are Chinese People 'Less Sensitive' About Privacy?" *Sixth Tone*.

Lieberthal, Kenneth. 1992. "Introduction: The 'Fragmented Authoritarianism' Model and its Limitations." In *Bureaucracy, Politics and Decision Making in Post-Mao China*, edited by Kenneth Lieberthal and David Lampton, 1–31. Berkeley and London: University of California Press.

Lieberthal, Kenneth, and David Lampton. 1992. *Bureaucracy, Politics, and Decision Making in Post-Mao China*. University of California Press.

Lieberthal, Kenneth, and Michel Oksenberg. 1988. *Policy Making in China*. Princeton University Press.

Liu, Yiling. 2019. "China's AI Dreams Aren't for Everyone." *Foreign Policy*. Available at https://foreignpolicy.com/2019/08/13/china-artificial-intelligence-dreams-arent-for-everyone-data-privacy-economic-inequality/. Accessed on 28 Feburary 2020.

Lv, Guang, and Linguo Hu. 2017. "3天变10分钟 广州商事登记实现全程"无人化" (From 3 days to 10 Minutes, Guangzhou Commercial Registration Has Achieved "Unmanned" Operation)." *Xinhua*.

Lynch, Marc. 2011. "After Egypt: The Limits and Promise of Online Challenges to the Authoritarian Arab State." *Perspectives on Politics* 9 (2): 301–310.

Ma, Jing, Yidan Luo, Jiaying You, and Wei Wei. 2018. "调查|谁在给你拨打骚扰电话? (Survey: Who Is Giving You Harassing Calls?)." *The Beijing News*.

Mabon, Simon. 2018. "Existential Threats and Regulating Life: Securitization in the Contemporary Middle East." *Global Discourse: An Interdisciplinary Journal of Current Affairs and Applied Contemporary Thought* 8 (1): 42–58.

MacFarquhar, Roderick. 1991. "The Anatomy of Collapse." *New York Reviews of Books*.

McKendrick, Kathleen. 2019. *Artificial Intelligence Prediction and Counterterrorism*. The Royal Institute of International Affairs (The Royal Institute of International Affairs).

McKinsey. 2017. *Jobs Lost, Jobs Gained: Workforce Transitions in a Time of Automation*. McKinsey Global Institute.

Merz, Fabien. 2019. "Europe and the Global AI Race." *CSS Analyses in Security Policy* 247.

Miller, Tom. 2017. *China's Asian Dream: Quiet Empire Building Along the New Silk Road*. London: Zed Books.

Montinola, Gabriella, Yingyi Qian, and Barry Weingast. 1996. "Federalism, Chinese Style: The Political Basis for Economic Success." *World Politics* 48 (1): 50–81.

Morozov, Evgeny. 2011. *The Net Delusion: The Dark Side of Internet Freedom*. Philadelphia, PA: Public Affairs Philadelphia.

Mozur, Paul. 2017. "Beijing Wants A.I. to Be Made in China by 2030." *The New York Times*.

Mozur, Paul, Raymond Zhong, and Aaron Krolik. 2020. "In Coronavirus Fight, China Gives Citizens a Color Code, With Red Flags." *The New York Times*.

Mueller-Kaler, Julian. 2020. *Europe's Third Way*. Atlantic Council. Available at https://www.atlanticcouncil.org/content-series/smart-partnerships/europes-third-way/. Accessed on 16 September 2021.

Nathan, Andrew. 2003. "Authoritarian Resilience." *Journal of Democracy* 14 (1): 6–17.

NISSTC. 2019. 人工智能安全标准化白皮书 (2019版) (2019 White Paper on AI Standardization). National Information Security Standardization Technical Committee. Available at https://www.tc260.org.cn/front/postDetail.html?id=20191031151659. Accessed on 21 June 2020.

Noesselt, Nele. 2014. "Microblogs and the Adaptation of the Chinese Party-State's Governance Strategy." *Governance* 27 (3): 449–468.

NSCAI. 2019. *The Interim Report of National Security Commission on Artificial Intelligence*. National Security Commission on Artificial Intelligence. Available at https://drive.google.com/a/nscai.org/file/d/153OrxnuGEjsUvlxWsFYauslwNeCEkvUb/view?usp=sharing. Accessed on 21 June 2020.

Nyman, Jonna. 2018. "Securitization." In *Security Studies: An Introduction*, edited by Paul Williams and Matt McDonald. London: Routledge.

Nyman, Jonna, and Jinghan Zeng. 2016. "Securitization in Chinese Climate and Energy Politics." *Wiley Interdisciplinary Reviews: Climate Change* 7 (2): 301–313.

Oxford. 2005. Artificial Intelligence. In *The Oxford Dictionary of Phrase and Fable*. Available at https://www.oxfordreference.com/view/10.1093/oi/authority.20110803095426960. Accessed on 18 June 2020: Oxford University Press.

O'Meara, Sarah. 2019. "AI Researchers in China Want to Keep the Global-Sharing Culture Alive." *Nature* 569.

Pang, Baoqing, Shu Keng, and Lingna Zhong. 2018. "Sprinting with Small Steps: China's Cadre Management and Authoritarian Resilience." *The China Journal* 80: 68–93.

Parris, Kristen. 1993. "Local Initative and National Reform: The Wenzhou Model of Development." *The China Quarterly* 134: 242–263.

Pei, Minxin. 2008. *China's Trapped Transition: The Limits of Developmental Autocracy*. Cambridge, MA: Harvard University Press.

Perry, Elizabeth. 2008. "Chinese Conceptions of "Rights": From Mencius to Mao—And Now." *Perspectives on Politics* 6: 137–147.

Pierskalla, Jan, and Florian Hollenbach. 2013. "Technology and Collective Action: The Effect of Cell Phone Coverage on Political Violence in Africa." *American Political Science Review* 107 (2): 207–224.

Qian, Yingyi, and Barry Weingast. 1995. "China's Transition to Markets: Market-Preserving Federalism, Chinese Style." *Journal of Policy Reform* (2): 149–185.

Qianzhan. 2018. 一文带你了解2018年全国各地人工智能行业最新政策! (One Article to Help You to Know Latest Regional AI Policies). Qianzhan Industry

Institute. Available at https://www.qianzhan.com/analyst/detail/220/180 329-8cef9d2f.html. Accessed on 29 Feburary 2020.

———. 2019. 2019年人工智能行业现状与发展趋势报告 *(2019 Report on Industry Status and Development Trends of AI)*. Qianzhan Institute. Available at https://bg.qianzhan.com/report/detail/1910081709070618.html. Accessed on 29 Feburary 2020.

Que, Tianshu, and Jiteng Zhang. 2020. "人工智能时代背景下的国家安全治理: 应用范式、风险识别与路径选择 (National Security Governance in the Era of Artificial Intelligence: Application Paradigm, Risk Identification and Path Selection)." 国际安全研究 *(Journal of International Security Studies)* 1: 4–38.

Ramanathan, Shriram. 2019. *China's Booming AI Industry: What You Need to Know*. Lux Research. Available at https://www.luxresearchinc.com/blog/chinas-booming-ai-industry-what-you-need-to-know. Accessed on 28 Feburary 2020.

Reardon, Sara. 2012. "Was the Arab Spring Really a Facebook Revolution?" *New Scientist*. Available at https://www.newscientist.com/article/mg21428596-400-was-the-arab-spring-really-a-facebook-revolution/. Accessed on 11 April 2016.

Ricker, Thomas. 2019. "The US, Like China, Has About One Surveillance Camera for Every Four People, Says Report." *The Verge*.

Roberts, Huw, Josh Cowls, Emmie Hine, Jessica Morley, and Mariarosaria Taddeo. 2020. "Governing Artificial Intelligence in China and the European Union: Comparing Aims and Promoting Ethical Outcomes." Available at SSRN: https://ssrn.com/abstract=3811034.

Roberts, Huw, Josh Cowls, Jessica Morley, Mariarosaria Taddeo, Vincent Wang, and Luciano Floridi. 2021. "The Chinese Approach to Artificial Intelligence: An Analysis of Policy, Ethics, and Regulation." *AI & Society* 36: 59–77.

Rod, Espen, and Nils Weidmann. 2015. "Empowering Activists or Autocrats? The Internet in Authoritarian Regimes." *Journal of Peace Research* 52 (3): 338–351.

Satariano, Adam. 2021. "Europe Proposes Strict Rules for Artificial Intelligence." *The New York Times*.

Schurmann, Franz. 1966. *Ideology and Organization in Communist China*. Berkeley and Los Angeles: University of California Press.

Shambaugh, David. 2001. "The Dynamics of Elite Politics During the Jiang Era." *The China Journal* (45).

Shambaugh, David L. 2008. *China's Communist Party: Atrophy and Adaptation*. Washington, DC: Berkeley.

Shan, Jie. 2017. "Tycoons Spark Discussion on Realization of Communism." *Global Times*.

Shanghai. 2017. 关于本市推动新一代人工智能发展的实施意见 *(Opinions on the Implementation of this City's New Generation of AI Development)*. Shanghai

Government. Available at http://www.shanghai.gov.cn/nw2/nw2314/nw2 319/nw12344/u26aw54186.html. Accessed on 29 Feburary 2020.

Shearlaw, Maeve. 2016. "Egypt Five Years on: Was It Ever a 'Social Media Revolution'?" *The Guardian*.

Sheehan, Matt. 2018. "How China's Massive AI Plan Actually Works." *Macro Polo*.

Sheehan, Matt. 2022. *"China's New AI Governance Initiatives Shouldn't Be Ignored"*. Carnegie Endowment for International Peace. Available at https://carnegieendowment.org/2022/01/04/china-s-new-ai-gov ernance-initiatives-shouldn-t-be-ignored-pub-86127. Accessed on 14 March 2022.

Shirk, Susan. 2007. *China: Fragile Superpower: How China's Internal Politics Could Derail Its Peaceful Rise*. New York: Oxford University Press.

Shirky, Clay. 2008. *Here Comes Everybody: The Power of Organizing Without Organizations*. Allen Lane

Sina. 2017. "刘强东: 共产主义将在我们这代实现 公司全部国有化 (Liu Qiang-dong: Communism Will Be Realized in Our Generation, All Companies Will Be Nationalized)." *Sina*.

Singer, Natasha, and Sang-Hun Choe. 2020. "As Coronavirus Surveillance Escalates, Personal Privacy Plummets." *New York Times*.

Stars. 2019. *The Chinese Bubble in Artificial Intelligence Is Insane*. Stars Insights. Available at https://www.the-stars.ch/wp-content/uploads/2019/ 07/HUANG-Yuanpu_The-Chinese-Bubble-in-Artificial-Intelligence-is-Ins ane.pdf. Accessed on 29 Feburary 2020.

Su, Xiaobo. 2015. "Nontraditional Security and China's Transnational Narcotics Control in Northern Laos and Myanmar." *Political Geography* 48: 72–82.

Swanson, Ana, and Paul Mozur. 2019. "U.S. Blacklists 28 Chinese Entities Over Abuses in Xinjiang." *New York Times*.

Trombetta, Maria. 2019. "Securitization of Climate Change in China: Implications for Global Climate Governance." *China Quarterly of International Strategic Studies* 5 (1): 97–116.

UN. 2012. *Big Data for Development: Opportunities and Challenges*. Available at http://www.unglobalpulse.org/sites/default/files/BigDataforDevel opment-UNGlobalPulseJune2012.pdf. Accessed on 15 February 2016: UN Global Pulse.

US. 2019. *Artificial Intelligence for the American People*. The White House. Available at https://www.whitehouse.gov/ai/. Accessed on 3 January 2021.

Vincent, James. 2020. "Google Favors Temporary Facial Recognition Ban as Microsoft Pushes Back." *The Verge*.

Vuori, Juha. 2008. "Illocutionary Logic and Strands of Securitization: Applying the Theory of Securitization to the Study of Non-Democratic Political Orders." *European Journal of International Relations* 14 (1): 65–99.

———. 2011. *How to Do Security with Words: A Grammar of Securitisation in the People's Republic of China*. University of Turku.

Wagner, Kurt. 2018. "Mark Zuckerberg Says Breaking Up Facebook Would Pave the Way for China's Tech Companies to Dominate." *Vox Media*.

Wallace, Jeremy. 2016. "Juking the Stats? Authoritarian Information Problems in China." *British Journal of Political Science* 46 (1): 11–29.

Wallace, Nick, and Daniel Castro. 2018. *The Impact of the EU's New Data Protection Regulation on AI*. Center for Data Innovation. Available at https://www2.datainnovation.org/2018-impact-gdpr-ai.pdf. Accessed on 28 January 2021.

Wang, Shaoguang. 2009. "学习机制, 适应能力与中国模式 [Learning Mechanism, Adaptability, and Chinese Model]." 开放时代 *(Open Times)* (7).

Wang, Xinxi. 2018. 伪概念泛滥, 认知模糊, AI营销或需一场再教育 *(Pseudo-Concepts Are Proliferated and Cognition Is Fuzzy, AI Marketing May Need a Re-education)*. Iyiou. Available at https://www.iyiou.com/p/80682.html. Accessed on 29 Feburary 2020.

Wang, Yinggu. 2019. 面对技术与市场"双瓶颈", 中国人工智能企业准备迎接寒冬 *(Facing the "Double Bottlenecks" of Technology and Market, Chinese AI Enterprises Are Preparing for the Cold Winter)*. China Money Network. Available at https://www.zhongguojinrongtouziwang.com/2019/03/07/70351/. Accessed on 29 Feburary 2020.

Wang, Yuanfeng. 2016. "王元丰: 第四次工业革命真的来了?: (Wang Yuanfeng: Is the Fourth Industrial Revolution Coming?)." *Global Times*.

Wang, Zhengxu. 2005a. "Before the Emergence of Critical Citizens: Economic Development and Political Trust in China." *International Review of Sociology* 15 (1): 155–171.

———. 2005b. "Political trust in China : Forms and Causes." In *Legitimacy: Ambiguities of Political Success or Failure in East and Southeast Asia*, edited by Lynn White. World Scientific Pub Co Inc.

Wang, Zhiqiu. 2015. ""云上贵州"好处不只一点 打破壁垒共享互通 (The Benefit of Guizhou on Cloud Is More Than a Little)." *Guizhou Daily*.

Webb, Amy. 2019a. "Build Democracy into AI: Human-Centered Policy is Needed to Wrest Control from China, Tech Giants." *Politico*.

———. 2019b. *The Big Nine: How the Tech Titans and Their Thinking Machines Could Warp Humanity*. PublicAffairs.

Webster, Graham, Rogier Creemers, Paul Triolo, and Elsa Kania. 2017a. "China's Plan to 'Lead' in AI: Purpose, Prospects, and Problems." *New America*.

———. 2017b. "Full Translation: China's 'New Generation Artificial Intelligence Development Plan' (2017)." *New America*.

Weidmann, Nils. 2015. "Communication, Technology, and Political Conflict: Introduction to the Special Issue." *Journal of Peace Research* 52 (3): 263–268.

Wiewiórowski, Wojciech. 2020. *Artificial Intelligence, Data and Our Values—On the Path to the EU's Digital Future*. European Union. Available at https://edps.europa.eu/press-publications/press-news/blog/artificial-int elligence-data-and-our-values-path-eus-digital_en. Accessed on 3 January 2021.

Wilkinson, Claire. 2007. "The Copenhagen School on Tour in Kyrgyzstan: Is Securitization Theory Useable Outside Europe?" *Security Dialogue* 38 (1): 5–25.

Wood, Peter. 2016. "Chinese Perceptions of the 'Third Offset Strategy'." *China Brief* 16 (15): 1–3.

Wright, Nicholas. 2019. "Artificial Intelligence's Three Bundles of Challenges for the Global Order." In *Artificial Intelligence, China, Russia, and the Global Order*, edited by Nicholas Wright. Maxwell Air Force Base, AL: Air University Press.

Xi, Jinping. 2016. "习近平: 为建设世界科技强国而奋斗 (Xi Jinping: Strive to Build a World Technological Power)." *People's Daily*.

———. 2017. *The Chinese Communist Party's 19th Party Congress Report*.

Xiaoxia. 2019. "China Has 854 mln Internet Users: Report." *Xinhua*.

Xing, Yijun, Yipeng Liu, and Cooper Cary. 2018. "Local Government as Institutional Entrepreneur: Public–Private Collaborative Partnerships in Fostering Regional Entrepreneurship." *British Journal of Management* 29 (4): 670–690.

Xinhua. 2015. 习近平考察贵州: 贵州发展大数据确实有道理 (Xi Jinping Visited Guizhou: It Is Reasonable for Guizhou to Develop Big Data).

Xu, and Naomi. 2019. "The Trump Administration Blacklisted Chinese A.I. Startups. But That Might Not Slow Them Down." *Fortune*.

Xu, Chenggang. 2011. "The Fundamental Institutions of China's Reforms and Development." *Journal of Economic Literature* 49 (4): 1076–1151.

Yan, Xiaojun. 2011. "Regime Inclusion and the Resilience of Authoritarianism: The Local People's Political Consultative Conference in Post-Mao Chinese Politics." *The China Journal* (66): 53–75.

Yang, Fang. 2015. 李克强"大数据词典": 共享、开放、安全 (Li Keqiang's Big Data Dictionary: Share, Open and Security).

Yang, Qingqing. 2019. "新一批人工智能"国家队"亮相: 京东、华为、小米等10家企业入选 (A New Batch of AI "National Teams" Debut: JD.com, Huawei, Xiaomi and Other 10 Companies Selected)." *21jingji*.

Yang, Yuan, Yingzhi Yang, and Fei Ju. 2017. "China Seeks Glimpse of Citizens' Future with Crime-Predicting AI."

Yap, Chuin-Wei Yap, and Gillian Wong. 2015. "China Wants to Tap Big Data to Build a Bigger Brother." *The Wall Street Journal*. Available at http://blogs.wsj.com/chinarealtime/index.php?p=28059&preview=true. Accessed on 20 November 2015.

Ye, Qing. 2019. "人脸识别让寻亲不再是大海捞针 (Face Recognition Makes Searching for Relatives no Longer a Needle in a Haystack)." *Technology Daily.*

Yu, Yifan. 2019. "Why China's AI Players Are Struggling to Evolve Beyond surveillance." *Nikkei Asian Review.* Available at https://asia.nikkei.com/Spotlight/Cover-Story/Why-China-s-AI-players-are-struggling-to-evolve-beyond-surveillance. Accessed on 1 March 2020.

Zeng, Jinghan. 2015a. "Did Policy Experimentation in China Always Seek for Efficiency?—The Case of Wenzhou Financial Reform." *Journal of Contemporary China* 24 (92).

———. 2015b. *The Chinese Communist Party's Capacity to Rule: Ideology, Legitimacy and Party Cohesion.* Palgrave Macmillan.

———. 2016. "China's Date with Big Data: Will It Strengthen or Undermine the Authoritarian Rule?" *International Affairs* 92 (6): 1443–1462.

———. 2019a. "Chinese Views of Global Economic Governance." *Third World Quarterly* 40 (3): 578–594.

———. 2019b. "Narrating China's Belt and Road Initiative." *Global Policy* 10 (2): 207–216.

———. 2020. *Slogan Politics: Understanding Chinese Foreign Policy Concepts.* London: Palgrave Macmillan.

Zeng, Jinghan, and Shaun Breslin. 2016. "China's 'New Type of Great Power Relations': A G2 with Chinese Characteristics?" *International Affairs* 92 (4): 773–794.

Zeng, Jinghan, Tim Stevens, and Yaru Chen. 2017. "China's Solution to Global Cyber Governance: Unpacking the Domestic Discourse of 'Internet Sovereignty'." *Politics and Policy* 45 (3): 432–464.

Zhang, Jingya. 2015. "本市城区郊区城关探头全覆盖 (Probes Fully Cover Our City)." 北京晨报 *(Beijing Morning).*

Zhang, Tiwei. 2012. "中国不能错过"第三次工业革命" (China Can Not Miss "The Third Industrial Revolution")." 中国青年报 *(China Youth Daily).*

Zhang, Yan. 2019. "公安部：进一步推广人工智能脸部识别技术进行打拐 (Ministry of Public Security: To Further Promote AI Facial Recognition Technology for Abduction)." *China Daily.*

Zhao, Dingxin. 2009. "The Mandate of Heaven and Performance Legitimation in Historical and Contemporary China." *American Behavioral Scientist* 53 (3): 416–433.

Zhao, Xiaoyan. 2019a. "浙江成立全国首支政务人工智能训练师队伍 (Zhejiang Created the First Government Affairs Focused AI Trainer Team)." *China News (Zhejiang).*

Zhao, Yuhan. 2019b. 人工智能核心产业规模达570亿元 (Value of AI Core Industry Reaches 57 Billion Yuan). *Beijing Daily.* Available at http://bjrb.bjd.com.cn/html/2019-08/12/content_12059007.htm. Accessed on 29 Feburary 2020.

Zheng, Yongnian. 2007. *De Facto Federalism in China: Reforms and Dynamics of Central-Local Relations*. Singapore: World Scientifics.

Zhong, Raymond. 2020. "Trump's Latest Move Takes Straight Shot at Huawei's Business." *The New York Times*.

Zhongkegaofu. 2019. *2019年中国人工智能产业研究报告 (2019 Report on China's AI Industry)*. 中科高服 (Zhongkegaofu). Available at https://mp.weixin.qq.com/s/BGBZ-aqd_AMeuFUYbnAjFg. Accessed on 29 Feburary 2020.

Zhu, Min. 2017. "朱民: 人工智能世界由中美两国主导 需加强人才储备 (Zhu Min: The AI World Is Dominated by China and the United States, and the Talent Pool Needs to Be Strengthened)." *Sina*.